Praise for *The Loom of God*

"Pickover's lively, provocative travel guide takes readers into the fascinating realm of mystic math, from perfectly strange numbers to fractured geometries and other curious nooks and crannies of ancient worlds and modern times."

—Ivars Peterson, *Science News*

"Chock full of mathematical treats, **The Loom of God** *takes you on a trip which explores ideas in a totally non-threatening, enjoyable format. . . . A must for the I-hate-math person as well as the mathematical explorer."*

—Theoni Pappas, author of *The Joy of Mathematics*

"From Pythagoras and his strange beliefs ('Don't eat food that causes you to flatulate') to Goedel's mathematical proof of the existence of God, Pickover guides the reader through the history of metaphysical logic. . . . Playful and surreal, **The Loom of God** *is accessible to anyone who's mastered long division."*

—*Discover* magazine

"Thought-provoking, entertaining, eerie, magical, and lively *are words that describe* **The Loom of God***, which weaves number theory, geometry, and computer power into a beautiful tapestry of numerology, fractals, and a fantasy of our world's future. Distinctive . . . dazzling . . . unconventional . . . great for recreational enrichment reading."*

—*Mathematics Teacher*

"Without peer as an idea machine, Cliff Pickover proves equally adept at writing. **The Loom of God** *is a well-crafted piece of mathematical science fiction."*

—Charles Ashbacher, *Journal of Recreational Mathematics*

"In **The Loom of God***, Cliff Pickover, in his irrepressible style, frolics through a forest of mathematical curiosities and historical tidbits, all skillfully woven into a futuristic fantasy, leaving you to wonder where he learned all that."*

—Julien C. Sprott, Professor of Physics, University of Wisconsin–Madison

"This book mixes science and philosophy with a large dash of humor, a cupful of religion, a teaspoon of whimsy. . . . Pickover hops from Pythagoras, the Incas, Stonehenge, Kabala, chakras, St. Augustine, and fractals with abandon. . . . By the way, if you'd like to note it in your planner, mathematics tell us the End of the World is scheduled for August 15, 2126. Be there or be square."

—*American Reporter*

Praise for Clifford A. Pickover

"I can't imagine anybody whose mind won't be stretched by [Pickover's] books . . . "
—Arthur C. Clarke

"Bucky Fuller thought big, Arthur C. Clarke thinks big, but Cliff Pickover outdoes them both."
—WIRED

"Pickover inspires a new generation of da Vincis to build unknown flying machines and create new Mona Lisas."
—Christian Science Monitor

"Pickover has published nearly a book a year in which he stretches the limits of computers, art, and thought."
—Los Angeles Times

"In recent years, Pickover has taken up the helm once worn by Isaac Asimov as the most compelling popular explainer of cutting-edge scientific ideas."
—In Pittsburgh

"Pickover just seems to exist in more dimensions than the rest of us."
—Ian Stewart, Scientific American

"Pickover is van Leeuwenhoek's 20th century equivalent."
—OMNI

"A perpetual idea machine, Clifford Pickover is one of the most creative, original thinkers in the world today."
—Journal of Recreational Mathematics

"Clifford A. Pickover is the heir apparent to Carl Sagan: no one else does better popular science writing than Pickover."
—Robert J. Sawyer, Nebula Award winner, author of Calculating God

"Add two doses of Isaac Asimov, and one dose each of Martin Gardner and Carl Sagan, and you get Clifford Pickover, one of the most entertaining and thought-provoking writers of our time."
—Michael Shermer, Skeptic

THE LOOM OF GOD

Selected Works by Clifford A. Pickover

The Alien IQ Test
Archimedes to Hawking
A Beginner's Guide to Immortality
Black Holes: A Traveler's Guide
Calculus and Pizza
Chaos and Fractals
Chaos in Wonderland
Computers, Pattern, Chaos, and Beauty
Computers and the Imagination
Cryptorunes: Codes and Secret Writing
————Dreaming the Future
Future Health
Fractal Horizons: The Future Use of Fractals
Frontiers of Scientific Visualization
The Girl Who Gave Birth to Rabbits
The Heaven Virus
Keys to Infinity
The Mathematics of Oz
Mazes for the Mind: Computers and the Unexpected
The Möbius Strip
The Paradox of God and the Science of Omniscience
A Passion for Mathematics
The Pattern Book: Fractals, Art, and Nature
The Science of Aliens
Sex, Drugs, Einstein, and Elves
Spiral Symmetry (with Istvan Hargittai)
Strange Brains and Genius
The Stars of Heaven
Surfing Through Hyperspace
Time: A Traveler's Guide
Visions of the Future
Visualizing Biological Information
Wonders of Numbers
The Zen of Magic Squares, Circles, and Stars

THE LOOM OF GOD

Tapestries of Mathematics and Mysticism

CLIFFORD A. PICKOVER

數學
宇宙之機杼

Mathematics is the Loom
Upon which God Weaves
The Fabric of the Universe

STERLING

New York / London
www.sterlingpublishing.com

STERLING and the distinctive Sterling logo are registered trademarks of
Sterling Publishing Co., Inc.

Library of Congress Cataloging-in-Publication Data Available

2 4 6 8 10 9 7 5 3 1

Published in 2009 by Sterling Publishing Co., Inc.
387 Park Avenue South, New York, NY 10016
Originally published in 1997 in hardcover as *The Loom of God: Mathematical Tapestries at the Edge of Time*
by Plenum Trade, New York and London
© 1997 by Clifford A. Pickover
Distributed in Canada by Sterling Publishing
c/o Canadian Manda Group, 165 Dufferin Street
Toronto, Ontario, Canada M6K 3H6
Distributed in the United Kingdom by GMC Distribution Services
Castle Place, 166 High Street, Lewes, East Sussex, England BN7 1XU
Distributed in Australia by Capricorn Link (Australia) Pty. Ltd.
P.O. Box 704, Windsor, NSW 2756, Australia

Sterling ISBN 978-1-4027-6400-4

For information about custom editions, special sales, premium and
corporate purchases, please contact Sterling Special Sales
Department at 800-805-5489 or specialsales@sterlingpublishing.com.

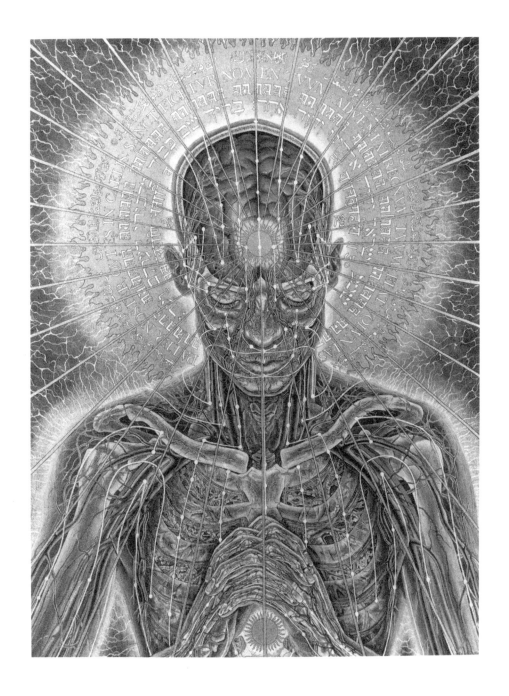

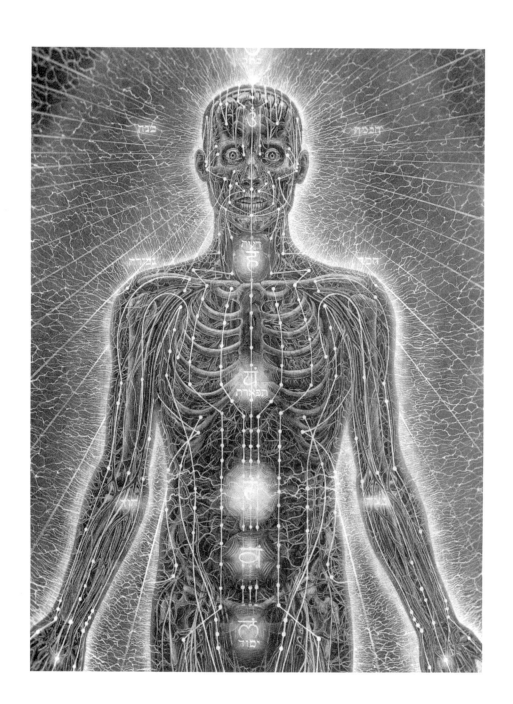

Acknowledgments

Some people think of God ... busily tallying the fall of every sparrow. Others—for example, Baruch Spinoza and Albert Einstein—considered God to be essentially the sum total of the physical laws which describe the universe.

—Carl Sagan, *Broca's Brain*

If we wish to understand the nature of the Universe, we have an inner hidden advantage: we are ourselves little portions of the universe and so carry the answer within us.

—Jacques Boivin, *The Single Heart Field Theory*

Science is not only compatible with spirituality, it is a profound source of spirituality.

—Carl Sagan, *The Demon-Haunted World*

I owe a special debt of gratitude to Linda Greenspan Regan, my editor at Plenum Publishing, for wonderful comments and suggestions. I also thank Arlin Anderson, Robert Stong, Clay Fried, Daniel Platt, Tassos Tzavaras, and Gary Adamson for useful comments and suggestions.

The frontispiece illustrations combining mysticism and the human body are from color originals appearing in *Sacred Mirrors: The Visionary Art of Alex Grey* (Inner Traditions International: Rochester, Vermont) and used with permission of Alex Grey. The first is titled "Praying," and includes prayers in different languages that encircle the head. The second is called "Psychic Energy System." In his amazing book, Alex Grey describes a vision of an intricate light pattern resembling a geometrical fiber pattern which he calls a Universal Mind Lattice. Alex and his wife Allyson note, "We both realized our vital connection with all beings and things in the Universe, with God. We felt that death was not to be feared because the Light was our spiritual core and we would eventually return to the profoundly transcendent bliss of this lattice realm."

I sent my epigram, "Mathematics is the loom upon which God weaves the fabric of the universe," to two individuals for calligraphic representation in Chinese and Farsi:

- The Chinese calligraphy on the title page was contributed by Dr. Siu-Leung Lee. Dr. Lee has been practicing the art of calligraphy for more than 40 years. Capable of writing in many styles, Dr. Lee has created his own style evolving from those of the Han and Jin dynasties. Besides Chinese calligraphy, Dr. Lee also plays the Chinese violin ("er-hu"), holds scientific patents, and owns a business. Dr. Lee was born in Hong Kong and resides in Columbus, Ohio. Literally translated, his calligraphy is: "Mathematics, The Loom of the Universe." (Here the word Universe can imply God.) The calligraphy makes use of a style that combines archaic structure and fluid movements to symbolize the dynamic nature of the universe.
- The Farsi script on page 7 was written by Jalil A. Taghizadeh of Sherman Oaks, California. Dr. Talat Bassari performed the translation from English to Farsi.

An excellent source of information about Pythagorean triangles and polygonal numbers is Albert Beiler's *Recreations in the Theory of Numbers* (Dover), and I recommend this book wholeheartedly.

Many of the old wood cuts appearing in *The Loom of God* are from the Dover Pictorial Archive.

The mathematical phenomenon always develops out of simple arithmetic, so useful in everyday life, out of numbers, those weapons of the gods: the gods are there, behind the wall, at play with numbers.

—Le Corbusier, *The Modulor*

All religions, arts, and sciences are branches of the same tree. All these aspirations are directed toward ennobling man's life, lifting it from the sphere of mere physical existence and leading the individual toward freedom.

—Albert Einstein

The end of all our explorations will be to come back to where we began and discover the place for the first time.

—T. S. Eliot

A world ends when its metaphor has died.
—Archibald MacLeish

Eternity is a child playing checkers.
—Heraclitus, 6th–5th century B.C.

Geometry aims at the eternal.
—Plato, 347 B.C.

Pythagoras

Contents

Introduction 15

Notes to the Paperback Edition 27

1 Are Numbers Gods? 29

2 The End of the World 42

3 Pentagonal Numbers 45

4 Doomsday: Friday, 13 November, A.D. 2026 60

5 666,666, Gnomons, and Oblong Numbers 70

6 St. Augustine Numbers 82

7 Perfection 86

8 Turks and Christians 99

9 The Ars Magna of Ramon Lull 104

10 Death Stars, a Prelude to August 21, 2126 122

11 Stonehenge   135

12 Urantia and 5,342,482,337,666   144

13 Fractals and God   149

14 Behold the Fractal Quipu   161

15 The Eye of God   167

16 Number Caves   176

17 Numerical Gargoyles   180

18 Astronomical Computers in Canchal de Mahoma   186

19 Kabala   190

20 Mathematical Proofs of God's Existence   198

21 Eschaton Now   207

22 Epilogue   225

Postscript 1. Goedel's Mathematical Proof of the Existence of God   227

Postscript 2. Mathematicians Who Were Religious   242

Postscript 3. Some Mind-Boggling Terminology   247

Author's Musings   251

Smorgasbord for Computer Junkies   257

Notes   270

References   275

Index   285

About the Author   288

Introduction

I have always thought it curious that, while most scientists claim to eschew religion, it actually dominates their thoughts more than it does the clergy.
—Astrophysicist Fred Hoyle

IS GOD A MATHEMATICIAN?

Mathematical inquiry lifts the human mind into closer proximity with the divine than is attainable through any other medium.
—Hermann Weyl (1885–1955)

Mathematics and mysticism have fascinated humanity since the dawn of civilization. Throughout history, numbers held certain powers that made it possible for mortals to seek help from spirits, perform witchcraft, and make prayers more potent. Numbers have been used to predict the end of the world, to raise the dead, to find love, and to prepare for war. Even today, serious mathematicians sometimes resort to mystical or religious reasoning when trying to convey the power of mathematics.

Has humanity's long-term fascination with mathematics arisen because the universe is constructed from a mathematical fabric? In 1623, Galileo Galilei echoed this belief by stating his credo: "Nature's great book is written in mathematical symbols." Plato's doctrine was that God is a geometer, and Sir James Jeans believed God experimented with arithmetic. Sir Isaac Newton supposed that the planets were originally thrown into orbit by God, but even after God decreed the law of gravitation, the planets required continual adjustments to their orbits.

Astronomy by Charles-Nicolas Cochin The Younger (1715–1790)

Is God a mathematician? Certainly, the world, the universe, and nature can be reliably understood using mathematics. Nature *is* mathematics. For example, the arrangement of seeds in a sunflower can be understood using Fibonacci numbers (1, 1, 2, 3, 5, 8, 13 …), named after the Italian merchant Leonardo Fibonacci of Pisa. Except for the first two numbers, every number in the sequence equals the sum of the two previous. Sunflower heads, like other flowers, contain two families of interlaced spirals—one winding clockwise, the other counterclockwise. The number of seeds and petals are almost always Fibonacci numbers.

The shape assumed by a delicate spider web suspended from fixed points, or the cross-section of sails bellying in the wind, is a catenary—a simple curve defined by a simple formula. Seashells, animal's horns, and the cochlea of the ear are logarithmic spirals which can be generated using a mathematical constant known as the golden ratio. Mountains and the branching patterns of blood vessels and plants are fractals, a class of shapes which exhibit similar structures at different magnifications. Einstein's $E = mc^2$ defines the fundamental relationship between energy and matter. And a few simple constants—the gravitational constant, Planck's constant, and the speed of light—control the destiny of the universe. I do not know if God is a mathematician, but mathematics is the loom upon which God weaves the fabric of the universe.

FILIX DENTATA

Fractal fern exhibiting self-similar branching, that is, similar structures repeated at different size scales. A single branch of the fern looks like a miniature copy of the entire fern.

MARILYN VOS SAVANT AND A UNIVERSE CALLED JUMBLE

Physicists are excited about discovering how reality behaves in terms of mathematical descriptions. This process is akin to discovering some hidden presence in the behavior of the universe—a gnosis. In this sense, physics is the inheritor of the tradition of Pythagoras.

—Anonymous IBM physicist

Marilyn vos Savant is listed in the *Guinness Book of World Records* as having the highest IQ in the world—an awe-inspiring 228. She is author of several delightful books and wife of Robert Jarvik, M.D., inventor of the Jarvik 7 artificial heart. Her column in *Parade* magazine is read by 70 million people every week. One of her readers once asked her, "Why does matter behave in a way that is describable by mathematics?" She replied:

> The classical Greeks were convinced that nature is mathematically designed, but judging from the burgeoning of mathematical applications, I'm beginning to think simply that mathematics can be invented to describe anything, and matter is no exception.

Marilyn vos Savant's response is certainly one with which many people would agree. However, the fact that reality can be described or approximated by *simple* mathematical expressions suggests to me that nature has mathematics at its core. Formulas like $E = mc^2$, $\vec{F} = m\vec{a}$, $1 + e^{i\pi} = 0$, and $\lambda = h/mv$ all boggle the mind with their compactness and profundity. $E = mc^2$ is Einstein's equation relating energy and mass. $\vec{F} = m\vec{a}$ is Newton's second law: force acting on a body is proportional to its mass and its acceleration. $1 + e^{i\pi} = 0$ is Euler's formula relating three fundamental mathematical terms: e, π, and i. The last equation, $\lambda = h/mv$, is de Broglie's wave equation indicating matter has both wave and particle characteristics. Here the Greek letter lambda (λ) is the wavelength of the wave-particle, and m is its mass. These examples are not meant to suggest that *all* phenomena, including subatomic phenomena, are described by simple-looking formulas; however, as scientists gain more fundamental understanding, they hope to simplify many of the more unwieldy formulas.

I side with both Martin Gardner and Rudolf Carnap who I interpret as saying: nature is almost always describable by simple formulas not because we have invented mathematics to do so but because of some hidden mathematical aspect of nature itself. For example, Martin Gardner in his classic 1985 essay "Order and Surprise" writes:

> If the cosmos were suddenly frozen, and all movement ceased, a survey of its structure would not reveal a random distribution of parts. Simple geometrical patterns, for example, would be found in profusion—from the spirals of galaxies to the hexagonal shapes of snow crystals. Set the clockwork going, and its parts move rhythmically to laws that often can be expressed by equations of surprising simplicity. And there is no logical or *a priori* reason why these things should be so.

Here Gardner suggests that simple mathematics govern nature from molecular to galactic scales.

Rudolf Carnap, an important 20th century philosopher of science, profoundly asserts:

> It is indeed a surprising and fortunate fact that nature can be expressed by relatively low-order mathematical functions.

To best understand Carnap's idea, consider the first great question of physics: "How do things move?" Imagine a universe called JUMBLE where Kepler looks up into the heavens and finds that most planetary orbits cannot be approximated by ellipses but by bizarre geometrical shapes that defy mathematical description. Imagine Newton dropping

an apple whose path requires a 100-term equation to describe. Luckily for us, we do not live in JUMBLE. Newton's apple is a symbol of both nature and simple arithmetic from which reality naturally evolves.

ARE MATHEMATICS AND RELIGION SEPARATE?

> *Had Newton not been steeped in alchemical and other magical learning, he would never have proposed forces of attraction and repulsion between bodies as the major feature of his physical system.*
>
> —John Henry, *Let Newton Be!*

In our modern era, God and mathematics are usually placed in totally separate arenas of human thought. But as this book will show, this has not always been the case, and even today many mathematicians find the exploration of mathematics akin to a spiritual journey. The line between religion and mathematics becomes indistinct. In the past, the intertwining of religion and mathematics has produced useful results and spurred new areas of scientific thought. Consider as just one small example numerical calendar systems first developed to keep track of religious rituals. Mathematics in turn has fed back and affected religion because mathematical reasoning and "proofs" have contributed to the development of theology.

In many ways, the mathematical quest to understand infinity parallels mystical attempts to understand God. Both religion and mathematics attempt to express relationships between humans, the universe, and infinity. Both have arcane symbols and rituals, and impenetrable language. Both exercise the deep recesses of our minds and stimulate our

Passion flower. The first Spaniards in South America connected this flower's structure with signs of Jesus' Crucifixion: The three upper parts of the pistil were the three nails. The five stamens were the five wounds surrounded by a crown of thorns. The ten "petals" represented the ten apostles of the Crucifixion.

imagination. Mathematicians, like priests, seek "ideal," immutable, nonmaterial truths and then often try to apply these truths in the real world. Some atheists claim another similarity: mathematics and religion are the most powerful evidence of the inventive genius of the human race.

Of course, there are also many *differences* between mathematics and religion. For example, many of religion's main propositions are impossible to prove, and religion often relies on faith unaffected by reason. In addition, while various religions *differ* in their beliefs, there is remarkable *agreement* among mathematicians. Philip Davis and Reuben Hersh in *The Mathematical Experience* suggest "all religions are equal because all are incapable of verification or justification." Similarly, certain valid branches of mathematics seem to yield contradictory or different results, and it seems that there is not always a "right" answer....

HOW MUCH MATHEMATICS CAN WE KNOW?

> *Einstein's fundamental insights of space/matter relations came out of philosophical musings about the nature of the universe, not from rational analysis of observational data—the logical analysis, prediction, and testing coming only after the formation of the creative hypotheses.*
>
> —R. H. Davis, *The Skeptical Inquirer*, 1995

We can hardly imagine a chimpanzee understanding the significance of prime numbers, yet the chimpanzee's genetic makeup differs from ours by only a few percentage points. These minuscule genetic differences in turn produce differences in our brains. Additional alterations of our brains would admit a variety of profound concepts to which we are now totally closed. What mathematics is lurking out there which we can never understand? How do our brains affect our ability to contemplate God? What new aspects of reality could we absorb with extra cerebrum tissue? And what exotic formulas could swim within the additional folds? Philosophers of the past have admitted that the human mind is unable to find answers to some of the most important questions, but these same philosophers rarely thought that our lack of knowledge was due to an organic deficiency shielding our psyches from higher knowledge.

If the yucca moth, with only a few ganglia for its brain, can recognize the geometry of the yucca flower from birth, how much of our mathematical capacity is hardwired into our convolutions of cortex? Obviously specific higher mathematics is not inborn, because acquired knowledge is not inherited, but our mathematical capacity *is* a function of our brain. There is an organic limit to our mathematical depth.

How much mathematics can we know? The body of mathematics has generally increased from ancient times, although this has not always been true. Mathematicians in Europe during the 1500s knew less than Grecian mathematicians at the time of Archimedes. However, since the 1500s humans have made tremendous excursions along the vast tapestry of mathematics. Today there are probably around 300,000 mathematical theorems proved each year.[1]

In the early 1900s, a great mathematician was expected to comprehend the whole of known mathematics. Mathematics was a shallow pool. Today the mathematical waters have grown so deep that a great mathematician can know only about 5% of the entire corpus. What will the future of mathematics be like as specialized mathematicians know more and more about less and less until they know everything about nothing?

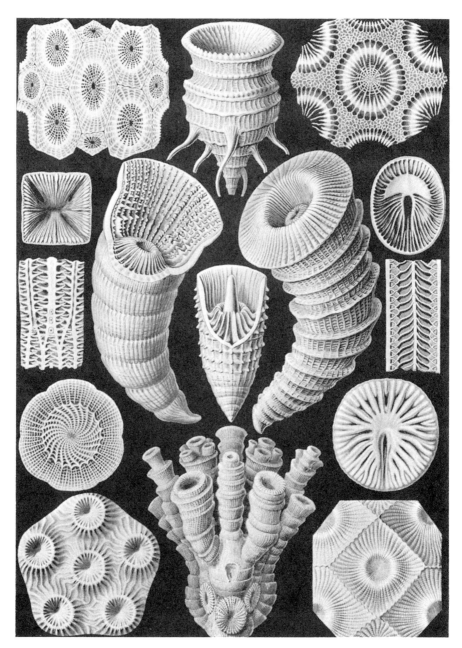

Various species of star corals exhibiting simple geometrical patterns.

More significant mathematical work has been done in the latter-half of this century than in all previous centuries combined.

—John Casti, *Five Golden Rules*

During their early days, both science and math have been connected with fictitious beliefs. Astronomy was connected with astrology, chemistry with alchemy, and mathematics with numerology. *The Loom of God* does not neglect this wild side of mathematics and its effect on human belief systems. The initial emphasis will be on Pythagoras, the ancient Greek mathematician whose ideas continue to thrive after three millennia of mathematical science. Philosopher Bertrand Russell once wrote that Pythagoras was one of the most intellectually important men who ever lived, both when he was wise and when he was unwise. Pythagoras was the most puzzling mathematician of history because he founded a numerical religion whose main tenets were transmigration of souls and the sinfulness of eating beans, along with a host of other odd rules and regulations. To the Pythagoreans, mathematics was an ecstatic revelation. They, like modern day fractalists, were akin to musicians. They created pattern and beauty as they discovered mathematical truths. Mathematical and theological blending began with Pythagoras, and eventually affected all religious philosophy in Greece, played a role in religion of

Engraving by French illustrator Gustave Dore (1832–1883) suggesting the mathematical fabric of reality.

the Middle Ages, and extended to Kant in modern times. Bertrand Russell felt that if it were not for Pythagoras, theologians would not have sought logical proofs of God and immortality.

HARMONY OF MATHEMATICS AND RELIGION

An intelligent observer seeing mathematicians at work might conclude that they are devotees of exotic sects, pursuers of esoteric keys to the universe.
—P. Davis and R. Hersh, *The Mathematical Experience*

Pure mathematics is religion.
—Friedrich von Hardenberg, circa 1801

The emphasis of this book is on *theomatics*—a word I coined in 1995 to denote the blending of mathematics and religion. I also discuss many end-of-the-world scenarios—

from ancient theological prophesies to modern astrophysical predictions. Numbers have played a central role in both religious and scientific apocalypses. To start you on your journey, we will first trace the logic of mathematics far back in time and examine humanity's search for ultimate answers to the mystery of existence, God, and the universe. The initial focus is around 550 B.C. because numbers had an auspicious reign in ancient Greece, especially for the Pythagoreans, the secret society devoted to exploring the mysteries of numbers. You will soon realize that both ancient and modern mathematicians trespass on territory that is often considered the exclusive province of religion.

There is a harmony in the universe that can be expressed by whole numbers. Numerical patterns describe the arrangement of florets in a daisy, the reproduction of rabbits, the orbit of the planets, the harmonies of music, the relationships between elements in a periodic table. On the controversial side, mathematics and religion have often come together to predict the end of the world, and numbers have been worshipped like gods. In this book, I'll give some unusual examples of this juxtaposition of God and mathematics, and also describe some of the current astronomical theories for the end of the earth.

Philosophers and writers make statements about mathematics that have religious undertones. For example, author Alan Watts has described mathematicians in the following way:

> The pure mathematician is much more of an artist than a scientist. He does not simply measure the world. He invents complex and playful patterns without the least regard for their practical applicability.

Through history, many philosophers and skeptics have probably made similar statements about religion.

Similarly, Aristotle describes mathematics in his *Metaphysics*: "Those who claim that the mathematics is not concerned with goodness and beauty miss the truth." Notice that if you were to examine the writings of many philosophers, and replace "mathematics" with the word "religion," their statements would be equally powerful and comprehensible. Why is this so? (For example, try this with Aristotle's quotation by replacing "mathematics" with "religion.") Is it because both mathematics and religion start with a belief (or axiom) system?

One of my favorite quotations describing the mystical side of science comes from Richard Power's *The Gold Bug Variations*:

Science is not about control. It is about cultivating a perpetual condition of wonder in the face of something that forever grows one step richer and subtler than our latest theory about it. It is about reverence, not mastery.

Again notice how the word "science" is easily replaced with "religion" or even "art."

The End of the World

We, while the stars from heaven shall fall,
And mountains are on mountains hurled,
Shall stand unmoved amidst them all,
And smile to see a burning world.
—Millerite Hymn, 1843

Throughout our history, various prophets of doom have predicted the end of the world using arcane mathematical manipulations. The end takes many forms: a huge comet crashing into the Earth, California sliding into the sea, the Apocalypse predicted in the Book of Revelation. No matter what form Doomsday takes, one thing is clear: the end of the world did not only intrigue ancient religious prophets; interest is still strong in our modern society. Just turn on your T.V. any Sunday morning to find some preacher telling you the world is about to end. Popular books predicting imminent disaster always find large and enthusiastic audiences. Today, in the United States there are probably more "doomists" than there ever were in some medieval or Roman town. Some fundamentalist Christians not only believe that there will be a Judgment Day when the world will end, but they also believe that the world *should* end.

The doomists have never been right—but one day they will be. Certainly the world will come to an end some time in the future, but more on this subject later...

Travel through Time and Space

There is no question about there being design in the Universe. The question is whether this design is imposed from the Outside or whether it is inherent in the physical laws governing the Universe. The next question is, of course, who or what made these physical laws?
—Ralph Estling, *The Skeptical Inquirer*, 1993

This book will allow you to travel through time and space, and you needn't be an expert in theology or mathematics. To facilitate your journey, I start most chapters with a dialogue between two quirky explorers who are interested in God and mathematics. You are Chief Historian of an intergalactic museum floating in outer space, a teacher and historian. Your able student is a scolex, a member of a race of creatures with bodies made of diamond.[2] Their hard bodies shield them from injury. Your personal scolex, Mr. Plex, helps you perform calculations and protects you from the dangers of time travel.

Prepare yourself for a strange journey as *The Loom of God* unlocks the doors of your imagination with thought-provoking mysteries, puzzles, and problems on topics ranging from Stonehenge to Armageddon. A resource for science fiction writers, a playground for computer hobbyists, an adventure and education for beginning students in theology, history, astronomy, and mathematics, each chapter is a world of paradox and mystery. Often various experiments in each chapter are accompanied by short listings of computer code in the Appendix. Computer hobbyists may use the code to explore a range of topics: from fractals, to asteroid cratering, to perfect numbers. However, the brief computer programs are just icing on the cake. Those of you *without* computers can still enjoy the journey and conduct a range of thought experiments. Readers of all ages can study theomatics using just a calculator.

As in all my previous books, you are encouraged to pick and choose from the smorgasbord of topics. Many of the chapters are brief and give you just a flavor of an application or method. Often, additional information can be found in the referenced publications. In order to encourage your involvement, I provide computational hints and recipes for producing the computer-drawn figures. For some of you, program code will clarify concepts.

Some information is repeated so that each chapter contains sufficient background information, but I suggest you read the chapters in order as you and Mr. Plex gradually build your knowledge. The basic philosophy of this book is that creative thinking is learned by experimenting.

An equation for me has no meaning unless it expresses a thought of God.
—Ramanujan (1887–1920)

Let me wrap up by mentioning some other topics you will encounter in this book as you journey from the ancient past to the far future. You'll meet enigmatic Greek warriors, Kabalists, St. Augustine, and Ramon Lull. You'll construct numerical gargoyles and visit prehistoric number caves. You'll hold an Incan fractal quipu in your hand, and discuss Doomsday. Strange numbers will surround you: pentagonal, perfect, oblong, and golden....

The oldest mathematical tablets found by archeologists date back to 2400 B.C., and I assume that humanity's urge to create and wonder about mathematics goes back to the earliest protohumans. Numbers were used by the ancient Amerindians, Sumerians, Babylonians, Chinese, Egyptians, and Indians. The Assyrians and Babylonians even assigned sacred deity-numbers to astronomical objects: our Moon was 30 and Venus was 15. Unfortunately, if we attempt to go back beyond the invention of writing, sometime around 6000 years ago in Sumeria, we find ourselves with little information. Mathematical and religious use of numbers before this time will forever remain a mystery.

In closing, let me remind readers that humans are a moment in astronomic time, a transient guest of the Earth. Our minds have not sufficiently evolved to comprehend all the mysteries of God and mathematics. Our brains, which evolved to make us run from lions on the African savannah, are not constructed to penetrate the infinite mathematical veil. And only a fool would try to compress several millennia's blending of mathematics and religion. We proceed.

Notes to the Paperback Edition

I am delighted that Sterling Publishing has invited me to prepare the new paperback edition of *The Loom of God*. Indeed, both public and scientific fascination in subjects relating to God, religion, mathematics, mysticism, and Doomsday seem to be growing, as indicated by the explosion of works ranging from the popular *Left Behind* series of apocalyptic novels and *The Da Vinci Code* to the recent movie version of *I Am Legend*.

Apocalyptic thinking about the End of Days for humanity is increasingly on our minds as we ponder the ramifications of catastrophic climate change, nuclear proliferation, and the likelihood of asteroid or comet impacts with Earth. Another form of the End of Days is the so-called technological singularity for which some futurists predict technological progress so astonishingly rapid that artificial intelligences will surpass us as the smartest and most capable life forms on Earth. Even the most serious cosmologists now ponder the ultimate fate of the universe. For example, if the acceleration of the universe continues as a result of a dark energy that seems to pervade our cosmos, this dark energy may eventually exterminate the universe in a "Big Rip" as all matter is torn apart.

With the year 2012 fast approaching, authors and publishers appear to be starting a countdown to another form of potential Doomsday apocalypse with a slew of books that predict calamity, spiritual transformation, or other momentous change in 2012—the year that a Mayan calendar cycle is completed. The Mayan Long Count calendar spans more than 5000 years, then resets to year zero in 2012. Depending on the author, we find any of the following are predicted to happen: an asteroid will collide with Earth, supervolcanoes erupt, the Earth's magnetic field reverses, or a mysterious global connection occurs with the emergence of a transhuman consciousness formed from the interactions of multiple human minds.

Our brains may be wired for belief in God or unseen entities. Religion is at the edge of the known and the unknown, poised on the fractal boundaries of psychology, history, philosophy, biology, and many other scientific disciplines. Because of this, the topics in *The Loom of God* are important subjects for contemplation. Even with the great scientific strides we will make in this century, we will nevertheless continue to swim in a sea of mystery. Humans need to make sense of the world and will surely continue to use both logic and religious thinking for that task. What patterns and connections will we see as the twenty-first century progresses? Who or what will be our God?

In closing, I would like to update readers on certain terms and concepts discussed in this book. Due to incredible advances in computing power, at least one numerical world record in this book has recently been broken. For example, as of 2008, the largest known perfect number (a concept discussed in Chapter 7) $2^{32,582,656} \times 2^{32,582,657} - 1$, with an amazing 19,616,714 digits.

As I mention in Chapter 14, the ancient Incas used *quipus* (pronounced "key-poos"), memory banks made of strings and knots, for storing numbers. Until recently, the oldest-known quipus dated from about 650 A.D. However, in 2005, a quipu from the Peruvian coastal city of Caral was dated to about 5000 years ago.

In Chapter 16, I mention that one of the world's fastest computers performs scientific calculations at a rate of around one teraflops—that is, one trillion floating point operations per second. In 2008, the fastest supercomputers achieve performances of over 1000 teraflops.

Chapter 21 discusses the infamous 1908 Tunguska explosion, which mysteriously leveled a very large area of Siberian forest. Supercomputer simulations performed in 2007 suggest that such devastation may have been caused by an impacting asteroid far smaller than previously thought, due to the asteroid's likely creation of a supersonic jet of expanding superheated gas. This finding suggests that any defensive strategy that we may develop to ward off asteroid impacts may need to consider the dangers of even smaller-sized menaces.

Although I did indeed "coin" the word *theomatics* with its particular meaning for use in this book, this word has been employed before my use. Prior use of this term often involved the numerological study of the Greek and Hebrew texts of the Christian Bible.

A great deal of this book focuses on Pythagoras, the ancient Greek mathematician and philosopher who is often credited with the invention of the Pythagorean theorem. However, evidence suggests that the theorem was developed by the Hindu mathematician Baudhayana centuries earlier, around 800 B.C., in his book *Baudhayana Sulba Sutra*. Pythagorean triangles were probably known even earlier to the Babylonians. Although the famous theorem that bears Pythagoras' name, $a^2 + b^2 = c^2$ for a right triangle with legs a and b and hypotenuse c, was probably known much earlier than the days of Pythagoras, some scholars have suggested that Pythagoras or his students were among the first Greeks to *prove* it.

For the latest reader comments, corrections, updates, and suggestions—including those with respect to the program code at the back of this book—please visit my new website: www.pickover.com/loom.html. As you peruse this book, I hope you will feel the sense of adventure I had when exploring all of the diverse subjects.

CHAPTER 1

Are Numbers Gods?

The hidden harmony is better than the obvious one.
　　　　　—Heraclitus, 6th–5th century B.C.

The year is 2080 and you are Chief Historian of an intergalactic museum
floating in outer space. Nicknamed *Theano*, your large ship carries objects of archeological
interest from several star systems. Currently on your view-screen is a marble temple
surrounded by massive fluted columns.

You turn to your assistant while pointing at the view-screen. "Mr. Plex, that beauty is
the Temple of Apollo at Corinth, circa 540 B.C."

Your assistant is a scolex, a member of a race of creatures with diamond-reinforced
exoskeletons that allow them to safely explore outer space.

"It's beautiful," Mr. Plex says. His slight hesitation is betrayed by the quivering of his
forelimbs. Perhaps he feels you are about to send him on a dangerous expedition.

You tap on the view-screen. "That's our destination, Mr. Plex. Now you know why
we've been studying the ancient Greek language. I want to understand humanity's use of
numbers in the search for ultimate answers. I want to find relationships between God and
mathematics. We'll see where it all started and how the world will end."

The scolex has a confused expression on his face. "Sir, how will the Temple of Apollo
help?" There is a vague metallic twang to his voice.

"First we're going to visit an ancient Greek religious cult that believed numbers were
sacred, godlike. Religion and math were one." You take a step closer to Mr. Plex.
"Interested in coming?"

The scolex takes a deep breath. "Sounds safer than your last idea of sending me into
space to study a black hole."

You raise your left eyebrow. "Certainly. What could go wrong?"

Mr. Plex points to the view-screen. "Why the Temple?"

"We're going to hide out in an unused backroom of the Temple. We'll set up our computers and observing equipment there." You gaze at the Temple and then turn back to Mr. Plex. "Ready?"

The scolex nods his large head and grins, revealing row upon row of diamond teeth that glitter in the incandescent light. "Ready!" he says.

"Calm yourself, Mr. Plex."

You press a button beneath the ship's view-screen and are transported to ancient Greece.

數 數 數 數

Mr. Plex quietly motions toward a row of olive trees. The ancient trunks are illuminated by a flood of moonlight, and their branches are cruelly twisted by the oceanic winds.

The weather is cool, but no more so than any bracing summer night on the spaceship *Theano*. In the distance you hear the gentle splashing of the Mediterranean.

What if you miscalculated and you are not in Pythagoras' time? What if you are miles from the designated location? You clear your mind of turbulent thoughts and envision the nearby sea crashing endlessly on the beach, one wave after another, like the beating of ancient drums.

You take a deep breath expecting the air of ancient Greece to be pure, invigorating. But instead you have to move your hand to cover your nose. You are not used to the pungent odor of rancid goat's milk.

Distant olive branches move. This must be the place. This must be the time. You tap Mr. Plex's forearm. "Follow me."

You move through olive trees that have branches so low and leafy you have to duck under them. There are places where someone could stand completely hidden by the trees.

Mr. Plex is so close you can hear him breathing. "Is that him?" he whispers as he points to your right.

You gaze between the olive branches and see the silhouette of a teenaged boy. He is smiling as he gazes upward at the Temple of Apollo. Or perhaps he is looking at the incredible lamp of stars.

"This can't be Pythagoras," you whisper to Mr. Plex.

There is a sound of a young woman laughing. Then in the darkness you see the dark form of a female adjusting her garments, whispering. The woman seems surrounded, cozily enclosed by the trees as their leaves flutter in the cool breeze. You smell her minty perfume.

You return your attention to the man and see a shiny discoloration on his thigh. Then without a word, he runs away along a stone path leading from the Temple of Apollo.

"Mr. Plex, there's got to be some mistake."

"What did you expect?"

"Not some smelly teenager. Not this."

"You expected someone more mythical, less human?"

"Yes. I expected a sage. A bearded wizard."

Mr. Plex tosses up his front limbs. "Ah, humans. You want to deify your heroes."

A storm seems to be coming from the west. The olive leaves are flapping like little birds, and some of the skeletal branches are scratching you. Even the wind seems ominous.

Strange odors waft up from the valley, and the wind periodically makes a sighing sound, like a dying god gasping for air.

Mr. Plex begins to shiver. "I hope we don't have to scrap the project, sir."

From a belt on your waist, you pull out a small hip flask of Kylonian brandy[1] and take a sip. "Care for a some?"

"Never touch the stuff, sir."

You perform a few mental calculations. "We've arrived too early. Let's jump ahead a dozen years. Give Pythagoras a chance to mature." You put away the flask. "Shall we give it another try?"

Mr. Plex nods.

You press a travel button on your belt. "Here we go-o-o...." You feel the usual jarring effect that accompanies this mode of time travel. You also feel faint nausea and smell the strange combined odor of vomit and rose blossoms. You and Mr. Plex are whisked away....

數 數 數 數

"I–I can't believe it," Mr. Plex screams. "That's him!"

You gaze at the view-screen hanging on a stone wall of the Temple of Apollo. You are in a cramped backroom which hasn't been used in the last few years, or so your staff has told you. You hope this is true, because you wouldn't want to be discovered with computer equipment. How would history be altered if the ancient Greeks gazed upon Mr. Plex and your computers?

A few dusty statues share the room with you. Perhaps they are gods which have fallen out of favor to make room for the latest Greek deities.

On the screen is a handsome, middle-aged man with long brown hair and beard. In the distance are some low hills.

"Pythagoras?" Mr. Plex asks you again.

Your heartbeat grows in intensity and frequency. "Yes," you whisper. "We got it right this time." You pause. "Pythagoras and his followers have some fascinating ideas and taboos. They're vegetarians, but never eat beans. They wear white clothes. They believe that numbers are divine ideas that create and maintain the universe."

Mr. Plex takes a step closer to the view-screen. "How are we observing them?"

"I released an electronic fly near Pythagoras' home. The fly has audio and video receptors."

"Can we turn up the volume?"

You nod and rotate a dial on the view-screen. Pythagoras' deep voice emanates from a cheap speaker dangling beneath the view-screen. For a moment you think you hear the chanting of monks.

"The study of arithmetic is the way to perfection," Pythagoras says to his disciples, a group of men and women clad in white. "By devotion to numbers, we discover both the divine plan and the mathematical rules of the universe."

Mr. Plex's eyes appear to be dilating with pleasure. "Mon Dieu," he whispers. "We're actually listening to the great Pythagoras."

"Quiet, Mr. Plex. I want to hear." You pause. "Look at what he's drawing."

Pythagoras sketches a figure with 10 dots forming a triangle:

Below the figure, Pythagoras writes: 1 + 2 + 3 + 4 = 10. Then he writes the Greek letters τετρακτυζ. The Pythagoreans gaze silently at the figure.

Mr. Plex looks at you. "Sir, what in the world is that?"

"It's the *tetraktys*." You pronounce it "te-tra-ktees" with an accent on the last syllable, your voice just a whisper. "They revere it. It's as important to them as the cross to the Christian. Initiates are required to swear a secret oath by the tetraktys when they begin their several years of silence as novices. They even pray to it."

"What's so special about a triangular array of dots?"

"They like the fact that it shows important musical intervals, 4:3 (the fourth), 3:2 (the fifth) and 2:1 (the octave). For example, choose any note. Another note an octave higher than this first note has a frequency twice that of the first note. I denote this by '2:1'. The musical interval called the fifth contains a note 3/2 times the frequency of the first note. And so on. Also notice that the first point in the tetraktys corresponds to the number of vertices of a point (1), the next pair to the vertices of a line (2), next a triangle (3) and at the bottom, a pyramid (4)." You pause to let the suspense heighten. "The most important thing is that 10 is a triangular number."

"A triangular number?"

"Yes." You drop your voice half an octave and assume a professorial demeanor. "Triangular numbers form a series, 1, 3, 6, 10,... corresponding to the number of points in ever-growing triangles." You take a piece of old charcoal from the Temple floor and draw on the wall:

"Incredible, sir. The possibilities are endless. The fourth triangular number is 10. I wonder what the 100th triangular number is?" He begins to count using his multiple limbs.

"Mr. Plex, there's an easier way. The *n*th triangular number is given by a simple formula:

$$\tfrac{1}{2}n(n + 1)$$

n is called the "index" of the formula. If you want the 100th triangular number, just use *n* = 100 for the index. You'll find that the answer is 5050."

Perhaps you detect admiration in Mr. Plex's eyes elicited by your mathematical prowess. "Sir, can we use a computer to determine the 36th triangular number?"

Next to you is a marble statue of Aphrodite. You reach into the statue's stomach where you have secretly stashed your notebook computer. A hinged door swings out, and you remove the computer and toss it to Mr. Plex.

Unfortunately, your aim is inaccurate, forcing Mr. Plex to make a leaping dive for the computer. He catches it but in doing so, crashes into a marble frieze running along the entablature decorated with marble representations of wild animals, centaurs, Hercules seizing Acheolus, and of men feasting. Hercules crashes down upon Mr. Plex.

Mr. Plex struggles to free himself of the horizontal Hercules and brushes himself off. "Never mind, sir. My diamond body can't be hurt by marble." He begins to furiously type on the computer's keyboard with his multiple legs. He hands you a computer printout:

Triangular Numbers:
1, 3, 6, 10, 15, 21, 28, 36, 45, 55, 66,
78, 91, 105, 120, 136, 153, 171, 190, 210, ...

"Sir, I can't believe it! The 36th triangular number is 666—the number of the beast in the Book of Revelation." Mr. Plex begins to quote from the Bible: '*Here is wisdom. Let him that hath understanding count the number of the beast; for it is the number of a man, and his number is six hundred, three score, and six.*' "

"Just coincidence, Mr. Plex."

"And the 666th triangular number is 222111. What a strange arrangement of digits!"

"Calm down, Mr. Plex. It's just coincidence."

"Sir, did you know that each square number is the sum of two successive triangular numbers?"

"What are you getting at?" Your voice is low.

"Square numbers are numbers like $5 \times 5 = 25$ or $4 \times 4 = 16$. Every time you add two successive triangular numbers, you get a square one. For example, $6 + 10 = 16$."

You are intimidated by Mr. Plex's mental agility, but then quickly snap back with a mathematical gem of your own: "Each odd square is 8 times a triangular number plus 1." You begin to draw a grid of squares on the wall. "Look at this." You point to the diagram.

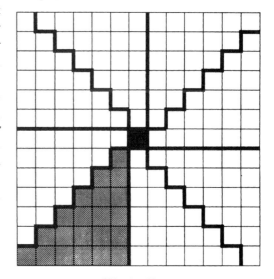

$$8T + 1 = K$$

You look back at Mr. Plex. "The Greek mathematician Diophantus, who lived 200 years after Pythagoras, found a simple connection between triangular numbers T and square numbers K. My diagram shows this. It has 169 square cells in an array. This represents the square number $K = 169$ (13×13). One dark square occupies the array's center, and the other 168 dots are grouped in 8 triangular numbers T in the shape of 8 right triangles. I've darkened one of the 8 right triangles."

Mr. Plex gasps, and you stare at each other. You feel as if the air has suddenly been evacuated from your dusty marble tomb.

"Sir," Mr. Plex whispers with a trace of hesitation, "no wonder the Pythagoreans worshipped triangular numbers. You can find an infinite number of triangular numbers such that when multiplied together form a square number. For example, for every triangular number, T_n, there are an infinite number of other triangular numbers, T_m, such that $T_n T_m$ is a square. For example, $T_3 \times T_{24} = 30^2$."

You slam your fist down feeling a slight pain as it makes contact with the cold marble.

You need to outdo Mr. Plex. After all, he is your pupil and assistant. You shout back, "666 and 3003 are palindromic triangular numbers. They read the same forwards and backwards."

Mr. Plex roars and begins to type on the notebook computer like a drunken spider. "It cannot be," he screams. "The 2662th triangular number is 3544453, so both the number and its index, 2662, are palindromic."

You feel a strange shiver go up your spine as you look into the scolex's glistening eyes. You feel a chill, an ambiguity, a creeping despair. Mr. Plex is still. No one moves. His eyes are bright, his smile relentless and practiced. Time seems to stop. For a moment, your secluded chamber in the Temple of Apollo at Corinth seems to fill with a cascade of mathematical symbols. But when you shake your head, the formulas are gone. Just a fragment from a dream. But the infuriating Mr. Plex remains.

"Mr. Plex, I grow weary of your little competition."

You gaze at the view-screen. The electronic fly is following Pythagoras as he walks to his domicile and prepares for bed.

"Come back," you say into a microphone to retrieve the fly.

"Sir, triangular numbers are fascinating. Are there other numbers like this? Pentagonal numbers? Hexagonal numbers? Did anyone worship these? What properties might these have?"

"Mr. Plex, that's the subject for another day. Throughout our journey I'll be emphasizing that humans have always intertwined numbers and religion, and simple formulas describe nature. It's as if God weaved the fabric of reality on a mathematical loom. But it's late now. Why don't we try to get some sleep?"

THE HISTORY AND SCIENCE BEHIND THE SCIENCE FICTION

> *It would not be right to say that the luxuriant development of religion in all its forms, a development that reached a kind of climax during the sixth century, helped science, nor yet that it harmed it. Then as now the two developments, scientific and religious, were parallel, contiguous, interrelated in many ways. They were not necessarily antagonistic.*
> —George Sarton, *A History of Science*

Pythagoras and Transmigration of Souls

The imagined monologue of Pythagoras is based on historical fact. Why is it that scientific and religious beliefs often took hold in the same minds through history? (For example, see *Postscript 2* at the end of this book which gives a listing of famous mathematicians who were religious.) Pythagoras is one of many mathematicians and scientists with *theomatic* tendencies. For example, German astronomer Johannes Kepler (1571–1630) was a mystic and theomatic. He published on the "music of the spheres" to explain why there were only a certain number of planets with particular orbits. He also felt that reality—from planetary motions to human illnesses—was an expression of God. If Kepler had a boil, he would not clean or lance it because he did not want to interfere with God's expression.

Another example of a theomatic was Isaac Newton. He published as much material on theology and alchemy as he did on physics. How did our modern science emerge from a tradition so rooted in mysticism and theology? How has theology changed as a result of the emergence of modern science?

Let's return our attention to ancient Greece where several communities arose sharing occult doctrines and various new revelations. Both men and women shared eschatological secrets in secret brotherhoods or communities of families. I believe that the most captivating of the ancient Greek societies were the Pythagoreans. Pythagoras was born on the island of Samos in 580 B.C. He lived roughly 80 years. In 525 B.C, Pythagoras traveled to Croton in southern Italy where he established secret societies devoted to exploring the mysteries of numbers.

Here are few Pythagorean beliefs, some of which had a lasting influence on humanity:

1   *Souls*—God created souls as spiritual entities which have the possibility of merging with the divine. The soul is an eternal, self-moving number which passes from body to body. Pythagoras believed in transmigration of the souls and claimed a semidivine status in close association with the god Apollo. He told his followers he remembered his earlier incarnations, thus increasing his prestige and the perception that he had superior knowledge of mathematical and spiritual lore. In fact, Pythagoras believed that the soul can leave the body either temporarily or permanently, and that it can inhabit the body of another human or even an animal. Like Jesus, the Pythagoreans lived simply and poorly, often going barefoot.
2   *Reality*—The universe was created and is currently guided by divine plan. The ultimate reality, however, is spiritual, not material, consisting of numbers and numerical relations. Ideas are divine concepts, superior to matter and independent of it.
3   *Life*—Marriage, faithfulness, and child-rearing are important. Children should be taught the faith in the power of numbers. Women are the equal of men. (Pythagoras had women followers.) The study of arithmetic is the way to perfection. By devotion to the sect and to numbers, individuals discover aspects of God's plan and the mathematical rules that guide the universe.

Pythagoras gives many rules to his followers. I have not come across rational explanations for many of the rules. For example, they must not eat meat, fish, or beans, or drink wine. The leaders must be celibate. They must avoid woolen clothes, and wear white clothes. (Wool is probably forbidden because it is an animal product. Linen, for example, is permitted.)

You've just read about some of Pythagoras' taboos, and you probably think that the Pythagoreans are a pretty unusual bunch. But wait a minute. They are even much stranger than you think, because the list of taboos multiplies. Pythagoreans are forbidden to:

1   Stir a fire with an iron poker.
2   Eat from a whole loaf of bread.
3   Pick up what has fallen.
4   Touch a white cock.
5   Urinate against the sun.

Some researchers have suggested that these and other Pythagorean taboos should not be interpreted literally. For example, "Do not urinate against the sun," might mean, "Be modest." If this is true, the inanity of certain phrases is certainly decreased.

All through the 4th and 5th centuries B.C., the Pythagoreans pursued their research, much of which was quite valid and important. How is it that such a weird bunch produced such fascinating mathematical ideas? One answer is that the fully initiated Pythagoreans weren't very concerned about taboos, and found themselves more interested in eschatology, science, and religion. Purification of the soul was accomplished by means of music and mental activity in order to reach higher incarnations. It is impossible to know about many of Pythagoras' doctrines because members were pledged to silence and secrecy. Sadly, after Pythagoras' death, many of his followers were massacred by the locals and, as with all religions, the martyrdom of his disciples increased Pythagoras' prestige.

Mathematics and Mysticism

Number is the measure of all things.
—Pythagoras

Speculations about numbers, numerical proportions, and harmonies are the most characteristic features of Pythagoreanism. Numbers resemble things and in some cases *are* things. The Pythagoreans even believed that abstract ideas have a number. For example, justice is associated with the number four and with a square. (Is this why today we have the expressions "a square deal" and "being all square?") Marriage is associated with the number five. Modern historians wonder what psychological associations the Pythagoreans used to make such numerical assignments.

Some numbers are compatible and friendly. Some are very evil, do not belong with other numbers, and bring calamity to humankind. Even numbers are female, odd numbers male. (Again we should ask what psychological associations Pythagoras made when coming to this conclusion. Are males more ornery and females more even-tempered?) The number 1 is the source of all numbers, and 2 the first female number. The first male number is 3, and the sum or union 2 + 3 = 5 stands for marriage. (Around 1000 B.C. in China, odd or male numbers were designated as white circles and even or female numbers as black.) The number 8 holds the secret of love since it adds male potency 3 to marriage 5!

Pythagoreans do not consider an odd composite number as properly masculine. (Composite numbers, as opposed to prime numbers, can be expressed as the product of other numbers, e.g., 10 = 5 × 2.) Mathematician Albert Beiler says, "True masculine qualities required a stern individuality such as only primes possessed." Therefore odd composites such as 9 or 15 are effeminate. But the number 3, an odd prime, is considered a fitting mate to feminine 2 to consummate marriage, 5. While males can unite with females, there is no barrier preventing males from coupling with

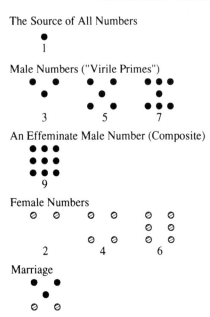

Numbers had sex according to Pythagoreans.

males or females with females. Pythagoreans believed it is the task of arithmetic to discover the relationships between numbers and their place in the divine plan. Composite numbers (e.g., 630) can be broken down into their various factors (e.g., $2 \times 3 \times 3 \times 5 \times 7$) for which we can find all sorts of relations. For Pythagoreanism, these relations are evidence that the Creator has a definite plan which can be discovered by a proper understanding of number sequences.

Later Pythagoreans assumed that the distances of the heavenly bodies from the Earth correspond to musical intervals. The Pythagoreans reasoned that just as musical harmonies can be described by certain ratios of string lengths, observation of the regular movements of planets and stars can be understood in numerical ratios.

It is often said that when Hippasus discovered that the ratio between the side and the diagonal of a rectangle cannot be expressed in integers, this shattered the Pythagorean world view. $\sqrt{2}$ is not an integer but rather 1.4142…, and this bothered Pythagoras a great deal. He loved the integers. The problem caused an existential crisis in ancient Greek mathematics, because $\sqrt{2}$ is the diagonal of a square with sides of length one, and it does not exist as a fraction. The digits of 1.4142… go on forever without any pattern. Pythagoreans dubbed these irrational numbers as *alogon* or unutterable.

Triangular Numbers

There was a theology of number, and the mathematician was a theologian who discovered and proclaimed the divine order.

—John McLeish, *The Story of Numbers*

We've seen that the sacred 10-dot triangle or tetraktys has a cosmic significance in Pythagoreanism. There is a relevant story of the merchant who asked Pythagoras what he could teach him:

"I will teach you to count," said Pythagoras.

"I know that already," replied the merchant.

"How do you count?" asked Pythagoras.

The merchant began, "One, two, three, four..."

"Stop," cried Pythagoras, "what you take to be four is ten, or a perfect triangle, and our symbol."

Pythagoras was most interested in numbers and relations that can be represented as dots drawn in sand, or as pebbles grouped in various ways. (Most historians believe that written numerals were not yet in use in Pythagoras' time.) As you and Mr. Plex observed, of particular interest to Pythagoras were the triangular numbers formed by triangular arrangements of dots. If you have visited a bowling alley, triangular numbers should be familiar: the 10 pins are set up in a triangular array. Pool players use a triangular frame to store 15 balls. 15 and 10 are triangular numbers. Pythagoras, as we've seen, knew that the fourth triangular number was ten and he called it the *tetraktys*. Pythagoreans took an oath to the tetraktys. It went something like this:

Bless us, divine number, thou who generates gods and men! O holy, holy tetraktys, thou containest the root and source of the eternally flowing creation!

This example is one of many in history where religion intermingles with mathematics. Why does this occur so often? Is it because both religion and mathematics strive to understand the ultimate answers? Their goals are often similar, but their methods very different.

Triangular numbers determined by $\frac{1}{2}n(n + 1)$ continue to fascinate mathematicians (but not the clergy) to this day.[2] Various beautiful, almost mystical, relations have been discovered. Here are just some of them:

- A number N is a triangular number if and only if it is the sum of the first M integers, for some integer M. For example, $6 = 1 + 2 + 3$.
- $T_{n+1}^2 - T_n^2 = (n + 1)^3$, from which it follows that the sum of the first n cubes is the square of the nth triangular number. For example, the sum of the first four cubes is equal to the 4th triangular number: $1 + 8 + 27 + 64 = 100 = 10^2$
- The addition of triangular numbers yields many startling patterns:

$$T_1 + T_2 + T_3 = T_4$$
$$T_5 + T_6 + T_7 + T_8 = T_9 + T_{10}$$
$$T_{11} + T_{12} + T_{13} + T_{14} + T_{15} = T_{16} + T_{17} + T_{18}$$

- 15 and 21 is the smallest pair of triangular numbers whose sum and difference (6 and 36) are also triangular. The next such pair is 780 and 990, followed by 1,747,515 and 2,185,095.
- Every number is expressible as the sum of, at most, three triangular numbers. German mathematician and natural philosopher Karl Friedrich Gauss (1777–1855) kept a diary for most of his adult life. Perhaps his most famous diary entry, dated 7/10/1796, was the single line:

EYPHKA! $num = \Delta + \Delta + \Delta$

which signifies his discovery that every number is expressible as the sum of three triangular numbers. The figures on p. 39 illustrate additional features of triangular numbers.

Triangular Number Contests

If you square 6, you get 36, a triangular number. Are there any other numbers (not including 1) such that when squared yield a triangular number? It is rumored that the next such number contains 660 digits, and humans have never found a greater example. Can you? Why are these types of numbers so secretive in our number system?

Some numbers such as 36 are both triangular and square. The next such *triangular-square numbers* are 1225, 41,616, and 1,413,721. What is the largest such number you can find?

We can use a little trick for determining huge triangular-square numbers. $8T_n + 1$ is always a square number. If the triangular number is itself a square, then we have the equation $8x^2 + 1 = y^2$. The general formula for finding triangular-square numbers is $1/32((17 + 12\sqrt{2})^n + (17 - 12\sqrt{2})^n - 2)$.

Can any triangular number (not including 1) be a third, fourth, or fifth power?

Proof without Words: A Triangular Identity

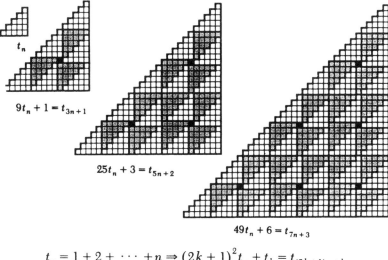

$$t_n = 1 + 2 + \cdots + n \Rightarrow (2k+1)^2 t_n + t_k = t_{(2k+1)n+k}$$

Proof without words: a triangular identity. These "visual proofs" illustrate various relationships between triangular numbers t_n. Small triangular numbers are represented by small triangles consisting of 15 squares. Larger triangular numbers are made by grouping these smaller triangular numbers, with the addition of individual squares as needed. (From Roger Nelsen; see References.)

$$\frac{1}{1} + \frac{1}{3} + \frac{1}{6} + \cdots + \frac{1}{\binom{n+1}{2}} + \cdots = 2$$

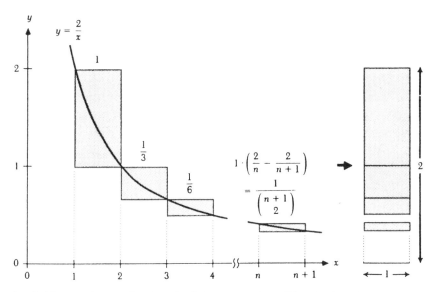

Proof without words: sum of reciprocals of triangular numbers. This "visual proof" illustrates the fact that the sum of reciprocals of triangular numbers equals 2. (From Roger Nelsen; see References.)

Mathematician Charles Trigg has found that $T_{1,111}$ and $T_{111,111}$ are 617,716 and 6,172,882,716, respectively. Notice that both the triangular numbers and their indices are palindromic, i.e., can be read backwards to yield the same number. Can you find a larger palindromic triangular number than these? Why the frequent occurrence of the digits 617 in these examples?

Obviously today we can compute huge triangular numbers using modern computers. What's the largest triangular number that Pythagoras could have computed? Would he have been interested in computing large triangular numbers?

If humanity devoted its energy to computing the largest possible triangular number within a year, how large a number would result? It turns out that this question has little meaning because we can construct arbitrarily large triangular numbers by adding zeros to 55 such as in 55, 5050, 500500, and 50005000. These are all triangular! Therefore, one large triangular number is:

50000000000000000000000000000000050000000000000000000000000000000000.

You can continue this pattern as long as you like. I wonder if Pythagoras or one of his contemporaries noticed a similar pattern.

Consider triangular numbers represented as an array of dots, as in the tetraktys. The dots are spaced 1/4-inch apart. What triangular number diagram has an edge length equal to the distance from our sun to the center of the galaxy (1.1×10^{21} feet)?

The longest lived stars will have used up all their fuel in 10^{12} years. If a computer computed successive triangular numbers at a rate of one a second, how large would be the triangular number when all the starts have died? What would be the triangle's size if drawn as an array of dots using a standard-sized font? Would it fit in the volume of the universe?

If you showed Pythagoras a huge triangular number produced using an average personal computer, and you convinced him of its validity, how would this have changed his philosophy and religion? How would Pythagoras' life have changed if he came into possession of a personal computer?

Religion and Odd Numbers

Pythagoras generally considered odd numbers good and associated with light, and even numbers evil and associated with darkness. Why is it that so many religions have unknowingly agreed with Pythagoras? For example:

> *Verily God is an odd number and loves the odd numbers.*
> —Islamic saying

> *The deity is pleased with the odd number [Numera deus impare gaudet]*
> —Virgil

> *There is no doubt that in the beginning the origin was one: the origin of all numbers is one and not two.*
> —Abdu'l-Baha, 19th Century

> *There is divinity in odd numbers.*
> —Shakespeare, *The Merry Wives of Windsor*

Here are some additional examples. The Talmud contains numerous examples of odd numbers and the avoidance of even ones. The Prophet Mohammad broke his fast by consuming an odd number of dates. When performing witchcraft, an odd number of persons should be present.

數 數 數 數

In this chapter, we've focused on number theory—the study of the properties of integers. Number theory is an ancient discipline. Much mysticism accompanied early treatises; for example, as we've seen, Pythagoreans based all events in the universe on whole numbers. Only a few hundred years ago courses in numerology were required by all college students, and even today such numbers as 13, 7, and 666 conjure up emotional reactions in many people. (Numerology is the study of mystical and religious properties of numbers.) Today integer arithmetic is important in a wide spectrum of human activities. It has repeatedly played a crucial role in the evolution of the natural sciences (for a description of the use of number theory in communications, computer science, cryptography, physics, biology, and art, see Schroeder (1984)).

CHAPTER 2

The End of the World

Truly the gods have not from the beginning revealed all things to mortals, but by long seeking, mortals make progress in discovery.

—Xenophanes of Colophon

The pure mathematician, like the musician, is a free creator of his world of ordered beauty.

—Bertrand Russell, *A History of Western Philosophy*

"Where the hell is Pythagoras?"

"Sir, our electronic fly indicates he's taking a nap beneath an old chariot in his back yard."

You remove a piece of yellowed paper from the marble floor of the temple. On the paper is some text beneath a woodcut showing apocalyptic horsemen flying through the sky on Judgement Day. (See p. 43)

"Ah, never mind Mr. Plex, we can use the time to examine a strange article one of my agents found in the year 1947." You pause. "There's always been predictions of the end of the world based on numbers. But predictions usually don't appear in serious mathematical journals." You raise your eyebrows. "This one appeared in a January 1947 issue of the *American Mathematical Monthly.*"

"Sir, let me see that," Mr. Plex says in a voice an octave too high.

Mr. Plex grabs the tattered paper from your hand and begins to read from the article:

The famous astrologer and numerologist Professor Umbugio predicts the end of the world in the year 2141. His prediction is based on profound mathematical and historical investigations. Professor Umbugio computed the value from the formula

$$W = 1492^n - 1770^n - 1863^n + 2141^n$$

for $n = 0, 1, 2, 3$, and so on up to 1945, and found that all numbers which he so obtained in many months of laborious computation are divisible by 1946. Now, the numbers 1492, 1770, and 1863 represent memorable dates: the discovery of the New World, the Boston

Massacre, and the Gettysburg Address. What important date may 2141 be? *That of the
end of the world, obviously.*

Mr. Plex lowers the slightly soiled slip of paper. "Sir, this is incredible. Could all the
numbers produced by the formula be divisible by 1946? Could it be that 2141 has anything
to do with the End of the World?"

You reach under a statue of Zeus and toss a notebook computer to Mr. Plex. "Write a
program, and see what numbers you get."

The Apocalyptic Horsemen. Woodcut by Hans Burgkmair, from *Das Neue Testament* printed by Silvan Othamar,
Augsburg, 1523.

Mr. Plex begins to type, and he soon hands you the results on a computer printout. The E symbols are the computer's way of representing scientific notation. For instance, 1.00E+02 would be another way of denoting 1.00×10^2 or 100.

```
 N   W
 1   0
 2   206276
 3   1.124106E+09
 4   4.106015E+12
 5   1.256519E+16
 6   3.478795E+19
 7   9.035302E+22
 8   2.246103E+26
 9   5.410357E+29
10   1.272996E+33
```

"Sir, the numbers grow awfully quickly! If the units were in years, the fifth value is larger than the number of years required for all the stars to have died out." Mr. Plex begins to pace. "How could scientists in the year 1946 determine that the results were all divisible by 1946? What is the W value for n = 100? Are the W numbers always divisible by 1946, or do they cease to have that property after n = 1945?"

"Mr. Plex—all very interesting unanswered questions. But they'll have to wait." You point to the-view screen and see Pythagoras rubbing his eyes and stretching. "Mr. Plex, our friend is waking up."

THE SCIENCE BEHIND THE SCIENCE FICTION

The "End of the World" formula really did appear in the following reference: Starke, E. (1947), Professor Umbigo's Prediction, *American Mathematical Monthly*, January, 54:43–44. I believe that all W numbers, even ones produced for n > 1945, are divisible by 1946. A detailed mathematical proof of this can be found in the *American Mathematical Monthly*. The proof relies on the fact that $x - y$ is a divisor of $x^n - y^n$ for n = 0,1,2, ….

The Appendix contains a BASIC program listing for computing Umbugio's End-of-the-World numbers.

CHAPTER 3

Pentagonal Numbers

I belong to the group of scientists who do not subscribe to a conventional religion but nevertheless deny that the universe is a purposeless accident. The physical universe is put together with an ingenuity so astonishing that I cannot accept it merely as a brute fact. There must be a deep level of explanation.

—Paul Davies, *The Mind of God*

Number systems are built according to different rhythms.
—Annemarie Schimmel, *The Mystery of Numbers*

On your view-screen is Pythagoras sitting in the sand beside one of his disciples. A light rain is falling on treeless hills. Pythagoras runs his hands through his wavy hair as he stares at the young man.

"Are you some kind of nut?" Pythagoras says.

The disciple stares at a diagram in the sand. "Oh great one, please explain further."

You are sitting beneath a dusty statue of Aphrodite. "Mr. Plex, can you get the electronic fly to go closer so we can see what Pythagoras is sketching?"

Mr. Plex goes to the view-screen and presses a few buttons. His multiple legs move in an oddball synchrony, giving him the appearance of a drunken spider. "That should do it, sir."

On the sand, Pythagoras has sketched a pentagonal arrangement of five dots.

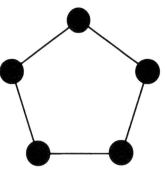

He begins to sketch other pentagons filled with 12 dots, and then 22 dots.

Pythagoras turns to the young man. "As with the triangular numbers, we can form successively larger pentagons by adding dots. The fifth pentagonal number is 35."

The disciple stares for a few seconds at a golden birthmark on Pythagoras' thigh, and then he looks up into Pythagoras' obsidian eyes.

The disciple says, "Let me determine the next pentagonal number." He makes pictures in the sand with a stick. A minute passes.

Pythagoras suddenly knocks the stick out of Heraclitus' hand. "Heraclitus, you're an oaf. The next pentagonal number is 51."

"Sir!"

Pythagoras looks down. "I'm sorry Heraclitus. I've had a lot on my mind lately. Problems at home. Please accept my apology for my rude behavior today."

Pythagoras pauses, and turns, and seems to be looking right at you through the view-screen. He reaches his hand toward you and says, "I hate flies."

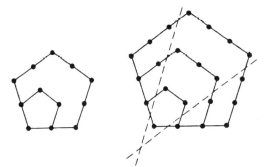

The view-screen goes dark.

You look at Mr. Plex. "No need to adjust the screen, Mr. Plex. He just squashed our electronic fly. We'll send out another."

"But sir, we'll never know how Pythagoras was able to compute pentagonal numbers with such agility."

"Mr. Plex, perhaps he knew of the formula $\frac{1}{2}n(3n-1)$. This gives the number of dots for the nth pentagonal number."

"Reminds me of the formula for triangular numbers we discussed last time." There is a sparkle in Mr. Plex's eyes. "This pentagonal formula generates 1, 5, 12, 22, 35 ..." Mr. Plex pauses. "Say, here's something weird."

"Mr. Plex, don't keep me in suspense."

"Well, for $n = 1$ we get a value of 1. For $n = 2$ we get 5. For $n = 3$ we get 12 ..., but the formula also produces values when n is negative." Mr. Plex begins to sketch on the marble temple wall:

```
Negative n   Positive n
... 40 26 15 7 2 0 1 5 12 22 35 51 ...
```

"Sir, what does it mean if the formula generates values for *negative* values of n? Can we draw some sort of antimatter pentagon?"

"Interesting, I have no idea." You are slightly intimidated by Mr. Plex's sage remarks and change the subject. "Mr. Plex, take a look at this."

You go over to the wall and start to sketch. "If you arrange all of the numbers you just wrote in ascending order, you get:

1 2 5 7 12 15 22 26 35 40 51 57 ...

"Now let's take successive differences and get:

```
1 2 5 7  12  15  22  26  35  40   51 57...
 1 3 2 5   3   7   4   9   5   11   6   (Differences)
```

"Mr. Plex, notice something interesting?"

"Mon Dieu!" the scolex screams. "If you skip every other difference you get the natural numbers, 1, 2, 3, 4, 5, 6, 7 ... and also the odd numbers, 1, 3, 5, 7, 9, 11, ..." Mr. Plex slowly looks up at the temple ceiling and whispers. "It almost makes me believe that a divine presence is in charge of all this."

"The ancients would agree with you. So would some modern mathematicians."

"Sir, can we experiment with other polygonal numbers—hexagonal, for example?"

"You bet, Mr. Plex. Amazingly we can compute all polygonal numbers with a simple formula

$$\frac{n}{2}[(n-1)r - 2(n-2)]$$

where r is the number of sides of the polygon and n is the 'rank'. The rank is simply an index which goes as $n = 1, 2, 3,...$"

You reach into a hidden compartment within a chipped statue of Poseidon, remove a notebook computer, and toss it to Mr. Plex. He types a short computer program. Out comes:

```
       Rank   1   2    3    4    5    6    7
Triangular    1   3    6   10   15   21   28
Square        1   4    9   16   25   36   49
Pentagonal    1   5   12   22   35   51   70
Hexagonal     1   6   15   28   45   66   91
Heptagonal    1   7   18   34   55   81  112
```

"Sir, what would 100-sided polygonal numbers be like?" There is awe in his voice.[1]

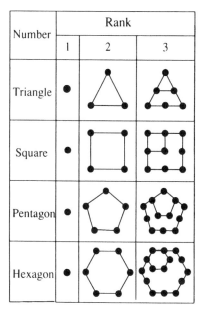

Polygonal number diagrams.

You raise both your eyebrows. You are eminently cool and in control. "As you add more sides to a polygon, the polygon becomes more circle-like." You pause to heighten Mr. Plex's suspense. "Mr. Plex, to take your question to the limit, what would circular, 'polygonal' numbers be like? Circles are infinitely sided polygons."

THE SCIENCE BEHIND THE SCIENCE FICTION

Nobody really knows why Hellenic philosophy developed as it did or why the Greeks proved themselves such expert philosophers.
—Peter Gorman, *Pythagoras: A Life*

Pentagonal numbers continue to amaze both professional and amateur mathematicians. For example, a very beautiful and important theorem involving pentagonal numbers was discovered by 18th-century mathematician Leonhard Euler. One day he started to multiply out the infinite product $(1 - x)(1 - x^2)(1 - x^3)(1 - x^4)$ ... and found that the first few terms were $1 - x - x^2 + x^5 + x^7 - x^{12} - x^{15}$ ... The exponents of x are pentagonal numbers! You can test this yourself by generating pentagonal numbers with $\frac{1}{2}n(3n - 1)$ for positive and negative values of n and arranging the resulting numbers in ascending order.

Do all numbers belong to more than one class of polygonal numbers? It turns out that there is a simple test for determining whether a number is a polygonal number for an n-sided polygon. Here is the secret recipe. Pick a number. Multiply it by $8(n - 2)$ and add $(n - 4)^2$ to the product. If the result is a square, then the number is a polygonal number for an n-sided polygon. For example, if we wish to ascertain if 50 is a pentagonal number, multiply by $8(5 - 2)$ giving 1200, and add $(5 - 4)^2$ giving 1201. Since there is no number when squared that yields 1201, 50 is not a pentagonal number.[2]

In the same way that any square number can be derived from the sum of two triangular numbers, a pentagonal number can be derived from the sum of the square of the same rank and the triangle of the preceding rank. For example, $T_2 + S_3 = P_3$, where T_2 is the second triangular number, and S_3 is the third square number, i.e. $3 + 9 = 12$. The figure below illustrates additional relationships for pentagonal numbers.

$$\frac{1 \cdot 2}{2} + \frac{2 \cdot 5}{2} + \frac{3 \cdot 8}{2} + \cdots + \frac{n(3n - 1)}{2} = \frac{n^2(n + 1)}{2}$$

Proof without words: sum of pentagonal numbers. This "visual proof" illustrates the summation of pentagonal numbers. The first row represents the pentagonal numbers: 1, 5, 12 ..., and the second row is another way of representing the same numbers. The objects can be placed together to form the staircase at bottom which contains $n^2(n + 1)/2$ boxes. (From William Miller; see References.)

As suggested in Chapter 1, so enamored were the Pythagoreans with whole numbers that irrational numbers such as $\sqrt{2} = 1.41 \ldots$ dealt them a blow from which they never recovered, and their serious study of mathematics gradually declined. As a *religion*, however, Pythagoreanism flourished in the form of a mystery cult through the Roman Empire. Some of its teachings can be traced to the beliefs of present-day occultists and theosophists.

Pythagoras' number theory was particularly influential because it played a major role in Plato's cosmology. After some time, Platoism and Pythagoreanism became so intertwined that today it is difficult to say whether a given philosopher owes more to Plato or to Pythagoras.

Incidentally, the character of Heraclitus encountered in this chapter actually did live within the period covered by Pythagoras' long life. Pythagoras' remorse for his rudeness is in keeping with what is known about his personality. Pythagoras is said to have been kind and never to have chastised anyone—not even a slave.

THE HISTORY BEHIND THE SCIENCE FICTION

The amount of eccentricity in a society has been proportional to the amount of genius, material vigor and moral courage which it contains.

—John Locke

Pythagoras: Thaumaturge and Premiere Kithara Player

Like many religious prophets, Pythagoras converted his followers by changing their inner thoughts. He held them spellbound—not by threats, criticism, or fear—but by his sheer intellectual power and mystical demeanor. Many historians have referred to Pythagoras as a *thaumaturge*, that is, one who performs wondrous things, a worker of miracles.

Unlike other religious teachers of the time (or other prominent religious prophets through history), Pythagoras had an advantage in being an expert musician who was said to control both animals and humans by playing his kithara (an ancient guitar). He often sang as he played.

There are many chronologies of Pythagoras' life, giving slightly different dates. Aristoxenus gives the following sequence of events: Pythagoras was born in 569 B.C., and spent his early years on Samos, a large Aegean island opposite the coast of Asia minor. (Samos was an important commercial center and panhellenic religious focus—perhaps due in part to its highly convoluted coastline with a large fractal dimension.) Samos' population was probably around 200,000.

Pythagoras was a wanderer. For example, he left Samos in 551 to go to Levant. He arrived in Egypt in 547 B.C. He was in Babylon from 525 to 513 B.C., and he returned to Samos in 513 B.C. In 510 he went to Croton, Italy, where his speeches won over more than 2000 people who became so excited that they did not resume their traditional family life at home. Instead, together with their wives and children, they established a Pythagorean school. All members shared their belongings. Pythagoras stayed in Italy for 39 years until his death in 470 B.C.

Anamnesis and the Number 216

Was Pythagoras once a plant? A seemingly bizarre question, but Pythagoras claimed he had been a plant and an animal in his past lives and, like Saint Francis, he preached to animals. Pythagoras and his followers believed in *anamnesis*, the recollection of one's previous incarnations. (Some historians suggest that Pythagoras promoted this belief to vindicate his superhuman powers.) During Pythagoras' time, most philosophers believed that only men could be happy. Pythagoras, on the other hand, believed in the happiness of plants, animals, and women.

In various ancient Greek writings, we are told the exact number of years between each of Pythagoras' incarnations: 216. Interestingly, Pythagoreans considered 216 to be a mystical number, because it is six cubed ($6 \times 6 \times 6$). (Six was also considered a "circular number" because its powers always ended in six.) The fetus was considered to have been formed after 216 days.

Pythagoras believed that even rocks possess a psychic existence. Mountains rose from the Earth because of growing pains of the Earth, and Pythagoras told his followers that earthquakes were caused by the shades of ghosts of the dead which created disturbances beneath the earth.

The number 216 continues to pop up in the most unlikely of places in theological literature. It is the famous and notorious number of Plato, occurring in an obscure passage from *The Republic* (viii, 546 B-D). It is associated with auspicious signs on Buddha's footprint.

The Infinite

"Whatever can be done once can always be repeated," begins Louise B. Young in *The Mystery of Matter* when describing our world. Similarly, Pythagoras believed that there were an infinite number of worlds in our universe, and that the worlds were a repetition of a finite quantity of matter. However, he also thought that infinity was the destroyer in the universe, the malevolent annihilator of worlds. If mathematics were war, the struggle was between the finite numbers and the concept of infinity. Pythagoras pitted the small numbers against the evil infinite, which he actually called an evil demon.

Gradually, the Pythagoreans became obsessed with infinity, and they concluded that numbers closest to one (and finiteness) were the most pure. Numbers beyond the range of ten were further from one and were less important.

The Greek philosopher Plato was also not comfortable with the concept of infinity, and he insisted that there was one universe bounded by a sphere which had nothing outside of it. Pythagoras, on the other hand, demanded the existence of the infinite. He admitted the existence of an infinite number of solar systems in a universe which was infinite.

Pythagoras told his followers that the universe was a living sphere whose center is the Earth. The Earth, too, was said to be a sphere, revolving like the planets. The Earth itself was an entire universe divided into five zones: Arctic, Antarctic, summer, winter, and equatorial.

Diogenes Laertius claimed that Pythagoras "was the first person to call the earth round, and to give the name of *kosmos* to the world."

The Secret Life of Numbers

Gods were numbers, pure and free from material change, and the worship of numbers one through ten was a kind of polytheism for the Pythagoreans. Pythagoreans believed that numbers were alive, independent of humans, but with a telepathic form of consciousness.

Humans could give up their three-dimensional life and telapathize with these number beings by using various forms of meditation. Meditation upon numbers was communing with the gods, gods who desired nothing from humans but their sincere admiration and contemplation. Meditation upon numbers was also a form of prayer which did not ask any favors from the gods.

These kinds of thoughts are not foreign to modern mathematicians who often debate whether mathematics is a creation of the human mind or exists in the universe independent of human thought. A few mathematicians even believe that mathematics is a form of human logic which is not necessarily valid in all parts of the universe.

The Number One and the Ship

Pythagoreans believed that our entire universe came into existence from numbers. The god One was the creator which produced Two, which in turn produced the first number Three, which is the symbol of the cosmos. Three also symbolized our three-dimensional world. Pythagoreans didn't speculate about higher-dimensional geometries and worlds because any theories about higher dimensions were considered impious and a study of evil since the infinite was identified with evil. The god One was also called "the cause of truth," "being," "the friend," and "the ship" because Pythagoreans thought of the cosmos as a ship whose keel was the central fire around which the planets revolved.

Pythagoras' Physical Deformity

Pythagoras had a golden birthmark on his thigh. Some scholars speculate that this may have discouraged him from accepting the Hellenic cult of the beautiful body and set him apart from others from an early age. On the other hand, as Peter Gorman points out (*Pythagoras: A Life*), Pythagoras made good use of his deformity by spreading a myth that the golden thigh indicated he was the son of Apollo or the god himself. Gorman writes:

> The golden thigh was certainly a birthmark, unless we accept the hypothesis that he had a golden plate transplanted into his flesh. His eccentric appearance and golden thigh both marked Pythagoras out as a freak who was bound to collide with the conformism of his native Samos.

Pythagoras is said to have shown his golden thigh to spectators at the Olympic games, causing a sensation and adding to his already growing fame. The birthmark convinced some that he was divine. Pythagoras was supposedly very fond of showing his deformity to all around him to impress them. When he showed the birthmark to the people of Croton in Italy, they screamed that he was the god Apollo incarnate. Eventually Apollo became the highest symbol of Pythagorean religious thought and was a personal guide to Pythagoras.

No Blood Sacrifice; Honey Allowed

Pythagoras believed in *theurgy*—sacrificial rituals during worship of the gods. However, Pythagoras prohibited blood sacrifice. This constraint led some of his later followers, like Empedokles of Akragas, to construct cow-shaped, sacrificial piñatas made of honey and barley which were offered on the altars of the gods.

Although Pythagoras believed that the gods should not be worshipped by blood sacrifices, he did not stress this idea so as to avoid offending his contemporary religious friends. For example, he suggested the more general prohibition that blood should not be shed in a temple. He also recommended that certain gods be worshipped with plants rather than animals—hence the offering of oak, laurel, or myrtle. Pythagoras was kind to animals because he believed they might contain the psyches of departed friends. In fact, Pythagoras once stopped a man from beating a dog, claiming to recognize the voice of a dead friend. His belief in reincarnation is also exemplified by an incident in a temple at Argos where he recognized the armor that he had worn in an ancient life.

Theurgy is a kind of magic which prepares the theurgist for contact with the gods. It is not a system of reward for sacrifices and prayers, an essential part of most revealed religions. As is true with all theurgists, Pythagoreans believed that the gods do not themselves answer prayers and sacrifices, nor do they help people merely because they worship properly and have good intentions. The theurgist believes that he literally forces the gods by magic to grant favors.

Although some historians report that Pythagoras joyfully sacrificed a hecatomb of oxen (a hundred animals) when he discovered his famous theorem about the right-angled triangle, this would have been scandalously un-Pythagorean and is probably not true. As noted, Pythagoras refused to sacrifice animals. Instead the Pythagoreans believed in the theurgic construction of agalmata—statues of gods consisting of herbs, incense, and metals to attract the cosmic forces.

The End of the World

The primary difference between Pythagoreanism and most other monotheistic religions is its idea that the cosmos is a beautiful harmony which is eternal. For example, Pythagoras believed that individual solar systems may be destroyed, but the universe as a whole will continue *ad infinitum*. The psyche still continues to be reincarnated every 216 years. Other religions suggest that the day of judgment will arrive when God overcomes evil, but this concept does not have a counterpart in Pythagoreanism.

With Pythagoreanism, there is no end to the universe. There is no final redemption.

Alien Abduction

UFOs and extraterrestrial life are hot topics today. But who would believe that these same ideas enthralled the Pythagoreans a few millennia ago? In fact, Pythagoreans believed that all the planets in the solar system were inhabited, and humans dwelling on Earth were less advanced than these other inhabitants. (The idea of advanced extraterrestrial neighbors curiously has continued, for some, to this day.) According to Pythagoreans, as one travels farther from the Earth, the beings on other planets and other solar systems become less flawed.

Pythagoras went as far as to suggest that the beings inhabiting the distant Milky Way were disembodied intelligences, mind-creatures with very tenuous physical bodies. Many of Pythagoras' followers believed that Pythagoras had once been a superior being who inhabited the Moon or the Sun.

The advanced creatures inhabiting regions beyond the earth could be visited if humans purified their psychic vehicle, which consisted of a mysterious substance similar to light. (On Earth this substance is mingled with other elements.) This is one of the reasons why Pythagoreans wanted to purify themselves—so that they could return to the upper regions of space where one attains divinity.

Pythagoreans further believed we possess an etheric body that shines within us. With the help of specific musical sounds, aromas, drugs, and incantations, the etheric body rides on beams of light toward the stars.

To Pythagoras, the whole universe was inhabited by beings, some with bodies made of fire and air and others who were totally devoid of matter. These superior beings sometimes guided humans.

Astraios from Outer-Space: Chthonic Power

In the last section, I mentioned how the Pythagoreans believed that the universe was inhabited by various kinds of beings. For example, they called one such superior space being *Astraios*. Some of these aliens were more physical than others and had bodies like ours. These aliens were darker than the star beings, had evil passions, and sometimes harmed humans. These beings followed dark gods. Pythagoras told his followers that these demons had to be propitiated, and should be appeased with "chthonic powers."

Astraios was among many of the Pythagorean space beings. He came to Earth riding on an arrow of some kind. Another interesting space being was Baris from the land of the Hyperboreans. (Hyperborea is a counter-earth or planet near the Earth. It blocks our view of a "central fire" about which the sun and planets revolve.)

Life in the Caves

> *One can assume that Pythagoras did cut his nails sometimes.*
> —Peter Gorman, *Pythagoras: A Life*

Our images of Greek heroes are usually ones of strength and beauty. This makes it all the more difficult to picture Pythagoras dirty, disheveled, and hungry, getting high on opium, and wallowing in bat guano in a damp cave. However while in Samos, around 520 B.C., Pythagoras withdrew into one of the many caves in the hills surrounding the city. This seemed to enhance the common impression that Pythagoras was separate from the rest of men. Notice that this self-imposed exile is also found in other religions. For example, John the Baptist and the Christians lived in caves.

Pythagoras reduced his hunger using a mixture of opium poppy seeds, sesame, and the bark of the squill (a Mediterranean herb that can act as a stimulant). The poppy causes a mystical state between waking and sleeping with bizarre visual and aural hallucinations. It was on one of these fantastic voyages of the mind that Pythagoras recollected his previous incarnations.

Incidentally, Pythagoras believed that people should not cut their nails or hair at religious festivals because these bodily growths were good. These growths were the gods' property and should not be cut. Pythagoras himself wore his hair long.

Beanophobia

During the creation of our world, beans were said to arise from the primordial slime at the same time as man. Perhaps this is one reason that Pythagoreans were forbidden to eat beans. (The literature on Pythagoras does not suggest explicitly that beans' gas-producing properties are the reason for the beanophobia.) Interestingly, Pythagoras was said to have been caught and slaughtered by the Kylonians (no relation to Klingons), when he stopped in front of a field of beans which barred his path of flight. Peter Gorman says that this story of Pythagoras' death should be dismissed as an absurd fiction.

Pythagoras' followers continued Pythagoras' aversion to beans. Empedokles, for example, said, "Wretches, you poor wretches, keep your hands off the beans."

Christianity and Pythagoreanism

There are startling similarities between early Christianity and Pythagoreanism:

- Both Christ and Pythagoras retreated into caves. In addition, like Christ, Pythagoras told his followers that they should love their enemies and prefer to suffer injustice rather than doing wrong. Also, Pythagoras could supposedly calm the waters and control the winds, another attribute of Christ.
- Once Pythagoras heard a loud, superhuman voice say, "Hail, Pythagoras," while crossing the river Kosas with some companions. This has certain biblical parallels with Jesus in the New Testament, for example when the heavens open, the Holy Spirit descends like a dove, and a voice says, "This is my beloved son ..."
- Pythagoras' miracle of the fishes seems to have much in common with the story told in John 21:6 and Luke 5:4. The miracle is described as follows. Pythagoras met fishermen on the beach and predicted the size of their catch, even specifying the number of fish. The fishermen were willing to do anything he told them if his prediction came true. When they brought in the next haul of fish, Pythagoras told the fishermen if his prediction was correct, he wanted them to let the fish go while they were still alive after they had counted them exactly. Pythagoras made a prediction and when the fishermen counted the fish, they found he had precisely guessed the number of fish. While the counting took place none of the fish remaining out of the water died during the time it took to count them. Pythagoras supervised the counting, then paid the fishermen the price of the fish. The fishermen spread the word about what happened, and Pythagoras became a demigod.

Pythagoras was once said to have simultaneously appeared in Croton and Metapontium. This was believed to result from his ability to walk on the water separating the two towns. A similar story is told of Christ.

What is a possible meaning of the similarities between Pythagoras and Jesus? It turns out that myths and histories are often transferred from one culture to another. For example, various stories in the book of Genesis reflect Babylonian mythology. The story of the flood has aspects of the epic of Gilgemesh. The biblical story of Gideon resembles Jason and the Argonauts in Greek mythology. Elements of the story of Sodom and Gomorrah have counterparts in Greek mythology. Apparently these stories were a kind of *lingua franca* (common language) in a time of oral tradition. (In fact, there were a dozen pre-Christian religions with "a savior who was crucified to atone for our sins.") The stories and histories shared features and were used to make points, sometimes by adding twists to familiar stories for the sake of impact or irony. Gradually, these ideas were folded into the written traditions. In the Bible, there are multiple occurrences of the same story in one sequence, and occasional contradictions between parallel telling of the same stories in different places.

Many great leaders have commonalities in their histories. Pythagoras and Jesus both had a circle, a "closed society" of disciples, but the same applies to Socrates and Plato. In many cases, oral memory exaggerates facts because disciples want their masters to be godlike. If you conduct your own research on Pythagoras, you should check historical sources because there are many neo-Pythagorean texts (especially from the 1st century) that are assigned to Pythagoras himself, but are not his actual words.

Frequently Asked Questions (FAQ)

Below is a list of frequently asked questions and answers regarding Pythagoras.

1   *Did Pythagoras marry?* Yes, he married the daughter of Brontinus of Croton, and he had several children including a daughter named Damo.
2   *How was Pythagoras' relationship with his daughter?* It was a good one. He entrusted Damo with his *Commentaries* and asked her to keep them secret from strangers. She must have been quite the devoted daughter, because she resisted disclosing the sect's secrets—and the temptation of selling his discourses for a great deal of money.
3   *What was his house like?* His house at Croton became a shrine and the object of superstitious awe. Reputedly a thief broke in one night, but dared not tell anybody the strange things he saw.
4   *Was Pythagoras, like Jesus, believed to have performed miracles?* Yes, most of the miracles attributed to Pythagoras occurred in Italy at the peak of his fame and activity. Here are some examples:
   a   He predicted the coming of a cargo ship carrying a dead body.
   b   He predicted the appearance of a polar bear in Kaulonia, an ancient Greek settlement in southern Italy.
   c   He told the Daunian bear to stop eating meat, and it obeyed. ("Daunian" refers to a region in southern Italy.)

d   He bit a poisonous snake to death in Etruria.

e   He persuaded a bull not to eat beans.

f   He predicted that there would be a revolution against the Pythagoreans.

5   *Was six a special number for the Pythagoreans?* Yes. Philolaus, an early follower of Pythagoras, thought six to be divine. It represented the six levels of existence from sperm to god. The first level of life is the organic and biological process of seed germination. The second level is plant life. The third level is comprised of animals. The fourth level is comprised of humans. The fifth is a race of demons who mediate between humans and god. The sixth level is comprised of the gods.

6   *What happened to a Pythagorean who blabbed secrets of the sect?* If a member broke the rules of the Pythagoreans by divulging its secrets, he was declared dead and expelled but not actually killed. A "cenotaph" (symbolic tomb) was erected for him.

7   *Are there more bizarre rules and taboos of Pythagoras?* Yes, the list seems endless:

a   Enter a temple wearing a white, clean cloak in which nobody has slept, because sleep is a sign of laziness.

b   Whenever there is a thunder-clap, touch the earth in memory of the creation of the universe.

c   Enter temples from the right-hand side.

d   Don't voluntarily shed blood in a temple. Instead one should sprinkle lustrations using sea-water.

e   Don't eat food that causes you to flatulate.

f   No women should give birth in a temple.

g   Don't burn bodies in a fire, because fire is divine and must not be dirtied.

h   Don't kill a flea in a temple, because the divine should not be saddled with such a measly body.

i   Don't clean your body or teeth with the leaves of cedar, laurel, cypress, oak, or myrtle.

j   Don't roast boiled food.

k   Don't pour libations with your eyes closed.

l   Don't sleep too much. Stay awake as long as possible to avoid the irrational state of sleeping. (One reason meat and beans were forbidden is because it was thought these foods encouraged drowsiness.)

m   Do not eat eggs.

n   Do not eat meat (i.e., "Beware of eating your ancestors.")

8   *Did Pythagoras follow his own rules?* Yes, as far as we know. He lived for the most part on bread and honey, with vegetables as dessert.

9   *What was initiation like for a beginning Pythagorean?* Those who wanted to become a follower of Pythagoras underwent an initiation whereby they could not speak out loud for five years. This means that they could not engage in business or even defend themselves in court. Only after the five years of silence was a student accepted as a full member of the Pythagorean society and permitted to see and study under Pythagoras. The scholars were accordingly divided into the

exoterici, or outer students, and *esterici*, or inner members who were able to share the great wisdom of Pythagoras himself.

10 *Was Pythagoreanism an instant hit?* Not always. When Pythagoras was around fifty-six years old and had returned to Samos, not many of the Samians had a desire for mathematical knowledge. Pythagoras, therefore, had to bribe a poor boy to be instructed in mathematics. However, eventually the boy became so fascinated by mathematics that he was willing to pay Pythagoras to teach him. The Pythagoreans were often ridiculed during the years stretching from the late 4th century B.C. to the 1st century B.C., and Pythagoras himself was treated as a charlatan who deceived his followers. Because the Pythagoreans were so active in local politics, the enraged members of the popular party of Croton burned down the house in which the Pythagoreans gathered. In one (unlikely) story of Pythagoras' death, the crowd followed him to a field of beans. As mentioned, he was said to have refused to tread on the beans and was killed. Another story has him escaping to Metapontum, where he abstained from food for forty days and starved himself to death.

11 *What is the counter-earth?* The Pythagoreans believed that there were ten planets which danced about a central fire—the Sun. To account for the fact that they only counted nine heavenly bodies in the solar system (including the Sun, Moon, and Earth), they invented a counter-earth which was supposed to be invisible.

12 *How could the gods understand humanity's prayers? Did the god's speak Greek?* The Greeks believed that the Greek language was the most tuned to the gods, and the utterances of gods, because the Greeks were the first humans to become aware of the existence of mathematical deities.

13 *What is a hermaphrodite number?* The number three is the first number, because one and two are the creators of numbers and not numbers themselves. Therefore the number one is potentially both an odd and even number, and called "hermaphroditic" or male-female (*arsenothelys*) by the Pythagoreans. One is the source of odd numbers. Two is the creator of even numbers.

14 *What was Pythagoras' concept of the soul?* Pythagoras believed that the soul was divided into three parts: feeling, intuition, and reason. Feeling was centered in the heart, intuition and reason in the brain. Animals and humans had feelings and intuitions. However, reason belonged to humans alone and was immortal. After death the soul underwent a period of cleansing in Hades, and then returned to Earth entering a new body in a chain of transmigration that could only be ended by a completely virtuous life.

Forty, Religion, Doomsday, Pythagoras

As mentioned in the previous section, one story of Pythagoras' death describes him escaping to Metapontum, where he abstained from food for 40 days and starved himself to death. Why is the number 40 so prominent in religion? The number 40 appears in the Old Testament in the story of Noah's ark where there are 40 days of rain. Christ spent 40 days in a desert where he was tempted by Satan. Ali Baba dealt with 40 thieves. Lent lasts for 40 days. There are 40 large stone pillars in Stonehenge arranged in a sacred circle (see Chapter 11). In Islamic tradition, the fetus is granted a soul after 3×40 days. The Children of Israel wandered in the desert for 40 years. Moses spent 40 days on the mountain. Christ rested for 40 hours in his tomb. Catholics have a ''40 Hours Devotion'' in which the sacrament is exposed for a period of 40 hours. St. Augustine interprets 40 as the product of 4 which signifies time, and 10 which means knowledge. Thus, 40 teaches us to live according to knowledge acquired during our lifetime. Both Jews (in the Talmud) and Catholics declare 40 as the ''canonical age'' of man—the age at which intellect is fully developed. Mohammad received his first revelation when he was about 40 years old. God kneaded Adam's clay for 40 days. At the resurrection, the skies will be covered with smoke and ash for 40 days. The resurrection lasts for 40 years.

Jews and Moslems believe that a woman should remain confined for 40 days after childbirth. The Islamic period of morning lasts for 40 days. Moslems believe that an animal should be served a special diet for 40 days before it is sacrificed. Moslems also believe that one should cut one's hair and nails once every 40 days.

In Moslem fables there are palaces with 40 columns. Mothers in Moslem fairytales produce 40 children in one birth. Heroes survive 40 adventures, kill 40 enemies, and find 40 treasures.

There are 40 martyrs mentioned in Islamic and Christian tradition (especially in the land of Anatolia). There were 40 brave men slain at the Prophet's tomb in Medina. Mohammad's son-in-law, the first Imam of Shiite Islam, had 40 disciples. There are 40 steps separating mortals from God. Muslims collect prophetic sayings in groups of 40. One such saying of the Prophet is: ''Whosoever among my people learns by heart 40 sayings (*hadith*) about religion will be resurrected at Doomsday.''

The Arabic philosopher Damiri claimed that if a blue-eyed child were suckled for 40 days by a black wet nurse, the child's eyes would turn black.

I can list dozens of additional examples of the number 40 in religion. Why so many 40s? And why should 40 appear in a description of Pythagoras' death? From a purely scientific standpoint, 40 was the number of days associated with the disappearance of the Pleiades, a conspicuous loose cluster of stars in the constellation Taurus consisting of six stars visible to the average eye. This 40-day period was known to the Babylonians and was used to time various festivals. However, I do not know if these facts relate to the frequent appearance of 40 in various religions.

Brief Summation, and a Prelude to Hypatia

The mathematical world of today owes Hypatia a great debt ... At the time of her death, she was the greatest mathematician then living in the Greco-Roman world, very likely the world as a whole.

—M. Deakin, *American Mathematical Monthly*, 1994

What became of Pythagorean thoughts and ideas? Pythagoras' philosophy, modified by Plato, outlasted all other philosophies of ancient Greece. Even up to the 6th century

A.D., the numerical gods were still worshipped, but during the Dark Ages their meanings were lost.

The Pythagoreans and their offshoot Platonists were the only ancient philosophical schools to allow women to share in the teaching. They were the only sects which produced outstanding women philosophers. Unfortunately, one of their best women philosophers, Hypatia of Alexandria (5th century A.D.), was martyred by being torn into shreds by a Christian mob—partly because she did not adhere to strict Christian principles. Interestingly, Hypatia is the first woman mathematician in the history of humanity of whom we have reasonably secure and detailed knowledge.

Hypatia

Hypatia was a respected, charismatic teacher, well liked by all her students. She was said to be physically attractive and determinedly celibate—to the point where she repelled one of her passionate admirers by shoving one of her used menstrual pads in his face and lecturing him on the shameful and dirty nature of what he thought was beautiful (the vagina).

Hypatia edited books on geometry, algebra, and astronomy. In one of her mathematical problems for her students, she asked them for the solution of the pair of simultaneous equations: $x - y = a$, $x^2 - y^2 = (x - y) + b$, where a and b are known. Can you find any values for x and y which make both of these formulas true?

Aside from being a mathematician, Hypatia assisted in the design of astrolabes, mechanical devices which replicate the motion of the planets. She also helped design urinonmeters to measure the specific gravity of urine. These were of potential use in determining the proper dosages of diuretics used to treat illnesses.

Sadly, we know more about Hypatia's death than other significant events in her life. The Christians were her strongest philosophical rivals, and they officially discouraged her teachings, which were Pythagorean in nature with a religious dimension.

In A.D. 415, a crowd of Christian zealots, led by Peter the Reader, seized her, stripped her, and proceeded to dismember her. They then burned pieces of her corpse. (Other accounts say she was burned alive, but these seem to be inaccurate.) Like many victims of terrorists today, she may have been seized merely because she was a well-known figure and prominent on the other side of the religious divide.

CHAPTER 4

Doomsday: Friday, 13 November, A.D. 2026

The 19th century Hasidic rabbi Menahem Mendelof Kotzk once asked some visiting scholars, "Where does God dwell?" They laughed at him and said, "God is everywhere, of course. The whole earth is full of his glory." The rabbi shook his head, then said, "God dwells wherever man lets him in."
—Stephen Mitchell, *The Enlightened Mind*

God is a fire, and we are all tiny flames; and when we die, those tiny flames go back in to the fire of God.

—Anne Rice, *Tale of the Body Thief*

"Sir, these things are starting to bother me."

You nod as you gaze up at a long marble frieze representing a battle, perhaps between Greeks and Trojans, with gods and goddesses looking on. Some of the most beautiful goddesses seem to be looking right at you as a blush of pure pleasure rises to your cheeks.

"What's the matter, Mr. Plex. Can't you appreciate beauty?" You look at Mr. Plex's brachiocephalic head and twitching diamond legs, and then say, "Never mind."

McDonald's hamburger wrappers from the year 2080 litter the marble floor like leaves after a hurricane. There is also a stench of rotting meat coming from somewhere in your close quarters. Perhaps it is time to do a little cleaning.

You stretch your legs. "Mr. Plex, let's go for a walk inside the Temple of Apollo. I'm getting a bit claustrophobic in this dump."

You get up, bang your head on an ankle of Aphrodite, walk to the door of the storage closet, and start to exit into the main chamber of the temple. "I'm glad we had our engineers design this door. It should keep out the curious."

Mr. Plex swivels his head 180 degrees. "Sir, aren't you worried we'll be seen?"

"No, it's early in the morning."

"What about our *Main Directive*? 'Time travelers must not interfere with the normal course of development of any civilization they might encounter.' "

"Mr. Plex, you worry too much. We're just taking a brief walk within the temple. Stretch our legs."

The cool breeze runs through your hair as you gather your thoughts. The air is spiced with scents of the sea. "Mr. Plex, most of the major religions give details on both the date and mechanism for Doomsday—the day when the world will come to an end. Theologians call it the *eschaton*, the divinely ordained climax of history."

The scolex is quiet for a few seconds as he scratches his head with his left foreleg. "Sir, I'm not very familiar with your ancient religions. Were the Christians the first to believe in the eschaton?"

"Don't think so. The expectation of an eschaton emerged in Judaism during Greek occupation. Perhaps Jews wanted to project the promise of justice into the next world." You pause and lean against a massive, fluted column. "Mr. Plex, also around that time were gnostic ideas which suggested humans were parts of God that had fragmented off, and

The Embarkation of Souls, by Gustave Dore. This haunting image reminds me of the statement from von Foerster's article: "Our great-great-grandchildren will not starve to death, they will be squeezed to death."

humans' goal was to find their true names and become part of God again. When no splinters of God remained, we would have the End of the World. Christianity seems to have inherited many of those ideas."

"Sir, I recall that your Earth had a number of messianic individuals in the 1990s warning of Doomsday. It was to occur in a convergence of environmental disasters, and only a select few would survive. For example, Luc Jouret—"

"Right, Luc Jouret was a homeopathic physician and spiritual explorer who died in a 1994 ritual. He was a leader of a cult which adopted many Christian ceremonies. Talked about the transformative power of fire: 'We are in the reign of fire. Everything is being consumed.' "

Mr. Plex shuffles closer to you as he gazes up at a statue of Athena and Hercules in a chariot. "Sir, Jouret predicted the end of the world was near. One of his followers wrote, 'We are leaving this earth to rediscover, lucidly and freely, a dimension of truth and absoluteness.' "

Mr. Plex appears to be deep in thought. His abdomen begins to pulsate.

You reach into your pocket. "Forget about Jouret. Here's an eschaton you can use your calculator to confirm." You bring out a few pieces of crumpled paper. "Our agents in the year 1960 have uncovered an article by serious mathematicians writing in the journal *Science*. The title is 'Doomsday: Friday, 13 November, A.D. 2026.' They claim that, on this date, human population will approach infinity—if it grows as it has grown in the last two millennia."

"Sir, this is a serious scientific paper?"

"*Science* magazine ain't no *National Enquirer*." You reach into your pocket to retrieve a piece of charcoal and begin to scrawl on Hercules' flat stomach. "The researchers' work contains the following formula for world population N as a function of time t

$$N = \frac{1.79 \times 10^{11}}{(2026.87 - t)^{.99}}$$

where time is measured in years A.D.[1] Just plug in a year t and you can calculate the population for that year. The researchers derive their model using a combination of empirical and theoretical reasoning dealing with fertility and mortality rates."

"Is it accurate?"

You nod. "Damn accurate. As they showed, the formula gave remarkably close figures for human population between the years 1750 and 1960. It was even in agreement with world population estimates when Christ was born!"

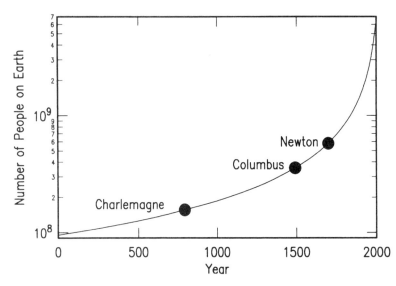

World population (*y*-axis) vs. year (*x*-axis). This curve is produced by the von Foerster Doomsday formula. The *y*-axis is logarithmic.

"Remarkable."

"There's just one problem." You take a step back and look directly into Mr. Plex's eyes. "In the year 2026 the U.S. population is infinite. N goes to infinity." You pause. "Some religious extremists have taken this to mean that in 2026 Armageddon comes— Doomsday. We all die."

A chill goes up your spine when you use the word "infinity."

Mr. Plex begins to choke. "Mon Dieu! How could that be?"

"No need to worry, Mr. Plex." You raise your left eyebrow and again stare directly into Mr. Plex's eyes. "We're from the year 2080, so obviously the formula doesn't work late into the 21st century—unless we changed something in our trip to Greece that will affect the destiny of the world."

Mr. Plex takes a step closer.

"Mr. Plex, let me quote from their paper where they describe a parameter in their model called t_0: 'For obvious reasons, t_0 [A.D. 2026] shall be called "doomsday," since it is on that date, $t = t_0$, that N goes to infinity and that the clever population annihilates itself.' "

You walk up to the seven-foot statue of Hercules, reach between his pectoral muscles, open a hinged door, and withdraw a notebook computer. You toss it to Mr. Plex.

Mr Plex's eyes bulge. "Sir, when did you put that there?"

"The notebook computer?"

Mr. Plex nods.

"I stashed it there one night. Thought we might need it."

Mr. Plex shrugs. "Let's program this on a computer."

Mr. Plex types in the formula, and together you watch as the numbers pour onto the computer screen:

```
Year A.D.   World Population
0           9.530058E+07
200         1.056239E+08
400         1.184714E+08
600         1.349001E+08
800         1.566543E+08
1000        1.868327E+08
1200        2.315209E+08
1400        3.045417E+08
1600        4.455130E+08
1800        8.329782E+08
1900        1.480906E+09
2000        6.884561E+09
2026        INFINITE!
```

Mr. Plex's abdomen is still pulsating. "Sir, this is incredibly accurate. But–but the population does go to infinity in the year 2026 as they predicted!"

You nod. "The constants in their formula, like 1.79×10^{11}, were derived from a study of available population data ranging over 100 generations from the time of Christ to the year 1958. However, their methods are so good that if Charlemagne had their initial equation $N = K/\tau^k$, and also several estimates of the world population available to him when he lived, he could have had predicted Doomsday accurately within 300 years. Elizabeth I of

England could have predicted the critical date within 110 years, and Napoleon with 30 years."

"Sir, what happens if we try to calculate the date when a hypothetical Adam emerged."

You shake your head. "If you use $N = 1$ the formula predicts Adam lived around 200 billion years ago. Obviously the Doomsday formula is not very accurate for these early times. After all, astronomers think the entire universe is less than 20 billion years old."

Mr. Plex gazes at you with admiration in his eyes. Your mathematical prowess never ceases to amaze and delight him.

You step away from the statue. "Various technological revolutions in human history show that food hasn't been a limiting factor to human growth. The Doomsday authors suggest our 'great-great-grandchildren will not starve to death, they will be squeezed to death.' "

You pick up your piece of charcoal and walk over to a nearby wall. "But you might be more interested in a beautiful equation published in a serious scientific journal in 1920. It was used to predict the United States population growth far into the future:

$$N = \frac{210}{1 + 51.5e^{-0.03t}}$$

N is the number of people in the United States in units of millions. e is Euler's constant and has a value of 2.71828.... t is time in years. The researchers confirmed that their magical equation was remarkably accurate from 1790 to 1910." You continue to draw on the wall:

```
Year   U.S. Population in Millions
1790   4
1850   23
1910   92
```

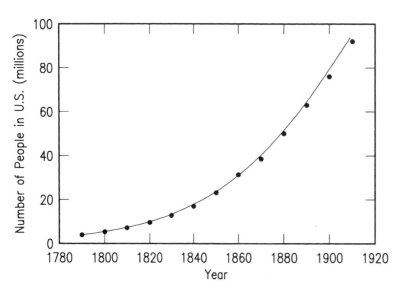

"Here are some census values for three years," you say pointing to the wall. "Years after the original paper publication, celebrations were held all over the world when various groups realized the accuracy of the United States population predicted by this formula. For example, thirty years after the equation was invented, the population in 1950 as predicted by the formula was 151 million, which was quite accurate."

Mr. Plex flexes his right forelimb. "Sir, I wonder if this formula also predicts a Doomsday when the population numbers explode and people die in

U.S. population (in millions) vs. year. The curve is predicted by the Pearl–Reed formula. The dots represent actual U.S. population values from census values. Notice the remarkably close fit of the prediction curve to the actual data.

utter horror, their compressed bodies hugging the coast of the U.S. like sardines in a tin can?"

"Quite poetic, Mr. Plex. Why don't you give it a try and see what numbers the formula yields."

Mr. Plex begins to type on the computer keyboard when suddenly you hear a voice.

A tall, attractive woman with long brown hair steps out from behind a column. You watch as she brushes back a lock of hair that the faint wind has caressed out of place.

She suddenly turns to you and Mr. Plex. "What in Hades is that?" she screams in Greek as she points to Mr. Plex.

"Pardon me, madam," Mr. Plex begins in rusty Greek, "but please speak slowly. We have only recently learned ancient Greek."

You slap his back. "Shut up you fool." A pain shoots up your arm. Diamond is very hard.

Mr. Plex drops the computer. You grab his forelimb.

"Quick, Mr. Plex, back to the storage closet. Maybe she won't realize we're hiding back there."

You run to the door to the backroom and notice that the woman has not followed. As you slam shut the door to the closet in the Temple of Apollo, you hear a masculine voice. Perhaps it is the woman's spouse or boyfriend.

"Theano," he calls to her. "What's wrong?"

There is no response from the woman. Perhaps Theano has fainted.

THE SCIENCE BEHIND THE SCIENCE FICTION

Apocalypse

Throughout history, both scientists and theologians have been interested in predicting the manner and date for the end of the world. For example, Nostradamus, the 16th century non-biblical prophet, predicted the end of the world to come around 1999.

Researcher Paul Bryant notes that some Christian theologians have suggested the world coming to an end in A.D. 2000. Interestingly, early church writers gave very specific support to this idea, and their argument is as follows. Rabbi Elias, Rabbi Ketina, Archbishop Ussher, Methodius, Lactantius, Irenaeus, and Barnabus each specifically state that the end of the age would come "6000 years" after Adam. Since the early Church belief held that Adam was created around 4000 B.C., this means that A.D. 2000 is the end of the world. For example, the (Apocryphal) Epistle of Barnabus says:

> And elsewhere he saith; If thy children shall keep my sabbaths, then I will put my mercy upon them. And even in the beginning of creation he makes mention of the sabbath. And God made in six days the works of his hands; and he finished them on the seventh day, and rested on the seventh day, and sanctified it. Consider, my children, what that signifies, he finished them in six days. The meaning of it is this; that in *six thousand* years the Lord God will bring all things to an end. For with him one day is a thousand years; as himself testifieth, saying, Behold this day shall be as a thousand years. Therefore, in six days, that is, in six thousand years, shall all things be accomplished. And what is that he saith, And he rested the seventh day: he meaneth this; that when his Son shall come, and abolish the season of the Wicked One, and judge the ungodly; and

shall change the sun and the moon, and the stars; then he shall gloriously rest in that seventh day. (Epistle of Barnabus, 13:3–6)

The Order of the Solar Temple

Luc Jouret, the apocalyptic leader from 1994 whom you and Mr. Plex discussed, was born in Zaire and went to Brussels for medical training. He studied acupuncture, homeopathy (a health treatment based on minute doses of medicine), the spiritual arcana of the Knights Templar (a mystical brotherhood banned in France in the 14th century), and the practices of the Reformed Order of the Temple (a 1990s group mixing Roman Catholicism, yoga, alchemy, and anticommunism). He eventually set up his own Geneva-based cult, the Order of the Solar Temple, which adopted many Christian rituals. Unfortunately, like David Koresh, he urged his followers to stockpile weapons. Fifty-three of his followers died by either murder or mass suicide, or some bizarre combination of the two. Two of his temples burned. Many of the bodies found were cloaked in ceremonial robes and arranged in a circle. His prediction of doom was true, at least, for his followers.

Doomsday Model

Scientists have always been interested in the problem of predicting the growth of a population. This is, of course, a very difficult problem which depends on a variety of parameters.

There really was a paper published in *Science* titled "Doomsday: Friday, 13 November, A.D. 2026." Despite the controversy this model caused, the von Foerster model has been a better predictor of world population growth than almost all other "accepted" models! Mathematical readers may be interested in some background. To derive their equation, the researchers assumed conditions on Earth which came close to being paradise: no environmental hazards, unlimited food supply, and the fate of a population as a whole determined by its fertility and mortality. With these conditions, the rate of change of N (the number of elements (people) in a population) is given by: $dN/dt = \gamma_0 N - \theta_0 N = a_0 N$ where $a_0 = \gamma_0 - \theta_0$ may be called the productivity of the individual element. The population displays a fertility of γ_0 offspring per unit time, and has a mortality of $\Theta_0 - 1/t_m$ derived from the life span for an individual element of t_m units of time. Integration of the dN/dt formula gives the well-known exponential growth of such a population with a time constant of $1/a_0$. In their article, the researchers describe a "people-stat" (like a thermostat) that would keep the world's population at a desired level. To keep our planet happy, they suggest that the people-stat should control the fertility γ to maintain it at the level $1/t_m$. This means cutting the birth rate to about half its present value or cutting the size of an average family to around two children.

The von Foerster paper stirred quite a controversy among scientists and laypeople. For example, researchers from Brookhaven National Laboratory wrote a letter to the *Science* editor stating that the population rate could never be as high as predicted by the von Foerster formula for latter years. They noted that the doubling time can never be much less than three-quarters of a year because, in human beings, about 270 days are required from conception to delivery. Another letter to the editor by W. Hutton stated: "It should be carefully noted that the limiting population may turn out to be larger than that estimated because people may become smaller. There is a solution that has not yet been suggested: cannibalism." Another letter to the editor noted that delays due to the age of puberty in a human female also limits the catastrophic prediction by the von Foerster formula.

Pearl-Reed Formulation

The second formula in this chapter

$$N = \frac{210}{1 + 51.5e^{-0.03t}}$$

was discussed in 1920 by R. Pearl and L. J. Reed (*Proceedings of the National Academy of Science*). (The Appendix contains a BASIC program listing for computing the U.S. population using the Pearl-Reed formula.) Interested readers may wish to see the derivation of this formula. The formula assumes that the time rate of change of N, the number of people in a population at time t, is a function only of N. This equates to: $dN/dt = F(N)$. The function $F(N)$ is called a growth function. If we assume that $N = 0$ is a solution to this differential equation (which means if no people are initially present, then no people will ever be present!), and if we assume that $N = C$ is a solution (which means that there is an upper limit, constant C, to the population which could occur because of limited food supplies), the simplest function $F(N)$ satisfying these assumptions is $F(N) = KN(C - N)$, where $K > 0$ for a growing population. Therefore, a differential equation for population growth is $dN/dt = KN(C - N)$, or $dN/dt = KCN - KN^2$. Here KN^2 is a death rate, and KCN is a birth rate. The solution to this differential solution satisfying the initial condition $N(t_0) = N_0$ is

$$N = \frac{C}{1 + (C/N_0 - 1)e^{-KC(t-t_0)}}$$

N approaches a limiting value C as t increases. N approaches zero as t approaches negative infinity.

The Pearl-Reed growth model for the U.S. population is equivalent to the growth model just derived. If t_0, t_1, and t_3 are equally spaced years, and x_0, x_1, and x_2 are the corresponding values for population, then the constants C and K are given by $C = 2x_1 - (x_0 + x_2)/(x_1^2 - x_0x_2)$ and $KC = [1/(t_1 - t_0)] \times [\ln(Cx_1 - 1)/(Cx_0 - 1)]$.

In 1920, Pearl and Reed gave another formula for the growth rate of the U.S. population: $y = 9,064,900 - 6,281,430x + 842,377x^2 + 19,829,500 \log x$, where x is the year, and y is the population. However, they note the following about this equation:

> How absurd this equation would be over a really long time range is shown if we attempt to calculate from it the probable population in, say, 3000 A.D. It gives a value of 11,822,000,000. But this is manifestly ridiculous; it would mean a population density of 6.2 persons per acre or 3968 persons per square mile. It would be the height of presumption to attempt to predict accurately the population a thousand years hence.

Using the first Pearl-Reed formula, $N = 210/(1 + 51.5e^{-0.03t})$, what U.S. population do you compute for the "Doomsday year" of 2026 predicted by the von Foerster formula in this chapter?

Another Population Model

You may be interested in a model of U.S. population growth I developed in the 1990s to estimate the U.S. population in millions changing through time. In the formula, N is the number people, and t is time. Just plug in a year t and you can calculate the population of people for that year:

$$N = 120 \times \ln[(1 + b)/(1 - b)] + 4$$

Here $b = (t - 1790)/250$. "ln" is the natural logarithm. The formula accurately predicts that in the year 1790 the U.S. population was only 4 million. More than two centuries later, in 1995, it predicts the U.S. population is 282 million. Pretty damn accurate. There's just one problem. Because $(1 - b)$ in the denominator is zero when $t = 2040$, in the year 2040 the U.S. population is infinite. N becomes infinite. Doomsday.

The Year of the Cusp, A.D. 2020

The power of population is indefinitely greater than the power in the Earth to produce subsistence for man. Population, when unchecked, increases in a geometrical ratio. A slight acquaintance with numbers will show the immensity of the first power in comparison of the second.

—Thomas R. Malthus, *An Essay on the Principle of Population*

Perhaps the most famous population prediction model comes from the prestigious Club of Rome, a gathering of international scientists and businessmen headed by Aurelio Peccei, former chief of Olivetti (business machines). The group used a computer to predict the changes in population, pollution, and natural resources through time. What they found was quite sobering. Their curves exhibited a maximum slope and ecological change in the year A.D. 2020, with eventual decay of the population due to starvation.

Years earlier, British economist Thomas Malthus was working on a related problem of how the breeding of species tended to use up resources. Rabbits, for example, bred until they used up their food supplies, and then they died of starvation. Then, more food grew. Then the rabbit populations grew. The Club of Rome predicted a "Malthusian Climax" of the population would take place in 2050 A.D. However, unlike the Malthusian cycles of growing and shrinking, this climax was a one-

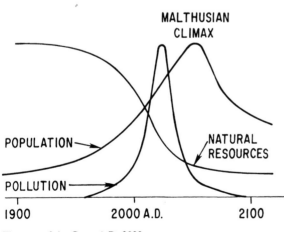

The year of the Cusp, A.D. 2020

shot deal. According to their computer model, humans depleted their natural resource, including minerals, and the population did not grow again. The declining population after 2050 A.D. was predicted to be horribly lacking in natural resources. Dennis Meadows, who led the MIT computer team, tried hundreds of different computer models. Unfortunately, all of their computer models of the future ended in calamity when a population growth factor was included. This would imply humanity is doomed, but the calculations don't take into account the possibility of an unexpected change in the world, or a powerful new idea that motivates poor people to control the number of children they have or that causes the rich to reduce their consumption. (I single out population control in the poor only because the wealthy usually have far fewer children than the poor.)

CHAPTER 5

666,666, Gnomons, and Oblong Numbers

There is no doubt that in the beginning the origin was one: the origin of all numbers is one and not two. Then it is evident that in the beginning matter was one, and that one matter appeared in different aspects in each element. Thus various forms were produced, and these various aspects as they were produced became permanent, and each element was specialized. Then these elements became composed, and organized and combined in infinite forms; or rather from the composition and combination of these elements innumerable beings appeared.

—Abdu'l-Baha, 19th century

"Mr. Plex, I think I'm in love."

There is a sudden crackling sound as Mr. Plex ambles over to you and his right foreleg crushes a half-eaten bag of tortilla chips he had brought back from the future. "Theano?"

"Right." You pause. "Let's take our minds off this."

"More Pythagorean mysticism?" Mr. Plex says eagerly.

You nod and walk over to a nearby wall of the backroom hideout. "As you know, Pythagoras' famous theorem is that in a right-angled triangle the sum of the squares of the shorter sides, a and b, is equal to the square of the hypotenuse c, that is $(c^2 = a^2 + b^2)$."

Mr. Plex nods.

"Mr. Plex, more proofs have been published of Pythagoras' theorem than of any other proposition in mathematics! There've been several hundred proofs."

"Sir, are Pythagorean triangles ones where a, b, and c are integers, like 3-4-5 and 5-12-13?"

"Correct, but Pythagoras' favorite, 3-4-5, has a number of properties not shared by other Pythagorean triangles, apart from its multiples such as 6-8-10."

"I know. It's the only Pythagorean triangle whose three sides are consecutive numbers."

"Very astute, Mr. Plex. It's also—"

Mr. Plex, beaming from your compliment, lifts his foreleg to silence you. "Sir, it's the only triangle of *any* shape with integer sides, the sum of whose sides (12) is equal to double its area (6)."

You continue, slightly annoyed by Mr. Plex's interruption. "It's truly an amazing triangle. But here's something that may make you think twice about 666, the Number of the Beast in the Book of Revelation."

"Go on, sir."

"There exists only one Pythagorean triangle except for the 3-4-5 triangle whose area is expressed by a single digit. It's the triangle 693-1924-2045. Its area is …" You pause to heighten the suspense. "666,666."

"Mon Dieu!" Mr. Plex screams like a raving lunatic. "It cannot be." His abdomen pulsates and his eyes quiver.

Again, for a moment, you think you hear the chanting of monks. It must be the wind.

You calmly reach for a notebook computer hidden beneath some paper napkins and soda cans. "Let me show you a magic set of formulas that will allow you to search for Pythagorean triangles. They've been known since the time of Diophantus and the early Greeks":

One Leg of Triangle: $X = m^2 - n^2$
Second Leg of Triangle: $Y = 2mn$
Hypotenuse of Triangle: $Z = m^2 + n^2$

"Sir, how do you use the formulas?"

"Just select any integers m and n and you should get a useful result."

"Fascinating, sir. Let me write a program to search for Pythagorean triplets."

Mr. Plex furiously types on the notebook computer and hands you a computer print-out:

X	Y	Z
4	3	5
6	8	10
8	15	17
10	24	26

"Mr. Plex. Here are some mind-boggling facts about Pythagorean triangles. In every triplet of integers for the sides of the triangles, one of them is always divisible by 3 and one by 5. The product of the two legs is always divisible by 12, and the product of all three sides is always divisible by 60." You pause. "Here's a star showing Pythagorean triangles each having one side equal to 120.

"Finally, Mr. Plex, can you find any triangles, like 3-4-5, that have consecutive leg lengths?"

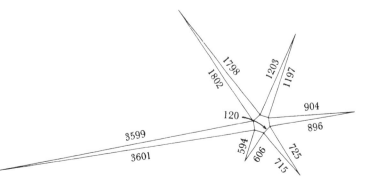

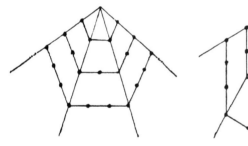

Mr. Plex seems breathless with your endless barrage of facts. "Do any other triangles like that exist?"

The view-screen, still hanging on the wall although slightly askew, flickers to life.

You walk over to the view-screen and tap it. "Mr. Plex, our new electronic fly must have reached its destination."

On the screen is Pythagoras talking to Heraclitus near a pool of water. The sky is reflected in the water; the clouds are bathing in its dark depths and trembling with the quivering eddies.

Pythagoras is drawing in the sand. He turns to Heraclitus. "If we start with one dot, we can continue to make squares by adding L-shaped arrangements of dots called *gnomons*." He pronounces it "nuh-muhns."

Pythagoras rests for a while, and you turn toward Mr. Plex. "The gnomon originally was simply a stick and its accompanying shadow when placed in the sun. Oenopides of Chios called a perpendicular a straight line drawn 'gnomon-wise' (κατα γνομονα). Theognis used the term gnomon to describe a carpenter's square. The Pythagoreans consider the odd numbers good because gnomons always form squares around odd numbers. Even numbers are evil because the sides of their gnomons are rectangular, not square."

Mr. Plex nods and turns his attention to the view-screen where Heraclitus is talking. "Pythagoras, how many dots are in each gnomon?"

"1, 3, 5, 7 … the odd numbers." He pauses. "If we generalize gnomons, all the polygonal numbers, like triangular and square numbers, are formed by adding different

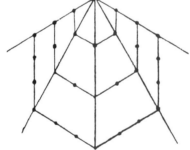

shaped gnomons successively to one. One is the first polygonal number of any form." Pythagoras draws in the sand. "Here is how we can construct pentagonal and hexagonal numbers by adding successive gnomons.

"In the case of the pentagon, the gnomons are 4, 7, 10 … with the common difference 3, and in the case of the hexagon we have 5, 9, 13 … with the common difference 4."

Heraclitus scratches his head. "Pythagoras, in general, the successive gnomonic numbers for any polygonal number, say of *n* sides, have $n - 2$ for their common difference."

Pythagoras nods. "Now let me tell you about the wonderful *oblong numbers*. They're formed by adding successive even numbers, and they're called oblong because they can be represented by oblong blocks of dots.

Pythagoras looks up to the heavens and says, "Let us pray for a moment." After a minute of silence, he says, "Just as the sum of any consecutive numbers (1, 2, 3, …) is a triangular number, the sum of any consecutive even numbers

2	6 (2+4)	12 (2+4+6)	20 (2+4+6+8)	30 (2+4+6+8+10)

is an oblong number with sides differing by 1. You can see this if you place successive gnomons in oblong numbers."

Heraclitus nods. "I notice that oblong numbers are exactly double the corresponding triangular numbers represented by two triangles side by side. Pythagoras, I think that any oblong number is the sum of two equal triangular numbers."

"Heraclitus, who's the teacher around here?"

"Sorry, Oh Great One."

Pythagoras looks up at some storm clouds. Already there is a faint drizzle. "Let us pray," he says.

While Pythagoras and Heraclitus are deep in prayer, you start discussing oblong numbers with Mr. Plex.

"Mr. Plex, these oblong numbers behave in some surprising ways. If you want to make a large oblong number, pick a large number, square it, and add it to itself. For example, $100^2 + 100$ would make an oblong number."

Mr. Plex is gazing at a nearby statue. Why is he ignoring you?

"Mr. Plex?"

"Sir, look here."

Your heart suddenly skips a beat. "What in the world?"

You go closer.

The statue of Aphrodite has changed in a subtle way. Most notably, she has six fingers on her right hand.

"Sir, was it like that before?"

"I don't think so."

"What could it mean?"

"We may be doing something here in ancient Greece, in this backroom of the temple, that is tearing the fabric of time." You gently touch Aphrodite's cold, marble hand. "Maybe our presence in the past is spawning a parallel universe subtly different from our own." You pause. "We've done something to alter the course of history, and maybe we're splitting off from our future universe."

You feel as if you are about to swoon as the rain upon Pythagoras faintly falls through the universe, upon all the living and the dead.

THE MATHEMATICS AND HISTORY BEHIND THE SCIENCE FICTION

The Praying Triangles

Pythagorean triangles with integral sides have been the subject of a huge amount of mathematical inquiry. For example, Albert Beiler, author of *Recreations in the Theory of Numbers*, has been interested in Pythagorean triangles with large consecutive leg values. These triangles are as rare as diamonds for small legs. Triangle 3-4-5 is the first of these exotic gems. The next such one is 21-20-29. The tenth such triangle is quite large: 27304197-27304196-38613965.

A few years ago I wrote a short story about these kinds of triangles. Before God created the Universe—eons before the Book of Genesis—He created mathematics. Since He is omnipotent, He created two classes of mathematical shapes: those that exist, and those that don't exist. In this latter group are such objects as planar triangles whose three angles' sum equals less than 180 degrees and Pythagorean triangles with two identical leg lengths. After a few centuries, these Pythagorean triangles prayed to God to make them exist so that they could have an effect on the real world. In God's infinite kindness He started to transform them in order to give them the legitimacy of existence for which they craved. However, due to distractions caused by a band of rowdy ellipses, God created Pythagorean triangles whose leg lengths differ by a mere one and forgot about the equal-legged triangles. To this day Pythagorean triangles with legs differing by one pray that their equal-legged brethren be transformed into "things that exist," but the fabric of the Universe has gelled, and God has since moved on to more weighty problems.

You can compute the Praying-Triangle leg lengths using the BASIC program listing in the Appendix. The recipe is as follows. Start with "1" and multiply by a constant $D = (\sqrt{2} + 1)^2 = 5.828427125\ldots$ Truncate the result to an integer value and multiply again by D. Continue this process for as long as you like creating a list of integers: 1, 5, 29.... To produce the leg length values for Praying Triangles, pick one of these integers, square it, divide by 2, and then take the square root. The two leg lengths are produced by rounding up and rounding down the result.

Divine Triangles

In 1643, French mathematician Pierre de Fermat wrote a letter to his colleague Marin Mersenne asking for a Pythagorean triangle the sum of whose legs and whose hypotenuse were squares. In other words, if the sides are labeled X, Y, and Z, this requires

$$X + Y = a^2$$
$$Z = b^2$$
$$X^2 + Y^2 = Z^2 = b^4$$

It is difficult to believe that the *smallest* three numbers satisfying these conditions are: $X = 4565486027761$, $Y = 1061652293520$, and $Z = 4687298610289$. I've called triangles of this rare type *divine triangles* because only a god could imagine a second solution to this problem. Why? It turns out that the second triangle would be so large that if its numbers were represented as feet, the triangle's legs would project from Earth to beyond the Sun!

If Pythagoras was told that a race of beings could compute the values for the sides of the second divine triangle, surely he would believe such beings were gods. Yet today we can compute such a triangle. We have become Pythagoras' gods. We have become gods through computers and mathematics.

Bricks Beyond Imagination

You and Mr. Plex discussed the interesting problem of finding Pythagorean triangles with integer values for the sides. A related, but fiendishly more difficult, task involves searching for solutions to the "integer brick problem." Here one must find the dimensions of a three-dimensional brick such that the distance between any two vertices is an integer. In other words, you must find integer values for a, b, and c (which represent the lengths of the brick's edges) that produce integer values for the various diagonals of each side: d, e, and f. In addition, the three-dimensional diagonal g spanning the brick must also be an integer. This means that the following equations must have an integer solution:

$$a^2 + b^2 = d^2$$
$$a^2 + c^2 = e^2$$
$$b^2 + c^2 = f^2$$
$$a^2 + b^2 + c^2 = g^2$$

No solution has been found. However, mathematicians haven't been able to prove that no solution exists. Many solutions have been found with only one non-integer side.

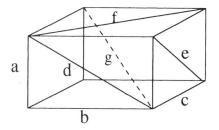

The Holy Grail

FACT FILE: 666

> Anonymous Interneter: *"What!? Does that mean if I buy 6 things on June 6, or look through a six-pane window from six feet away at six o'clock, I'm worshiping the devil?!"* A response: *"Quite simply, yes!"*

As discussed in this chapter, the number 666,666 has mathematical significance for the study of Pythagorean triangles. Here are some other (non-mathematical) 666 anecdotes.

- On July 10, 1991, Procter & Gamble announced that it was redesigning its moon and stars company logo. The company said it eliminated the curly hairs in the man in the moon's beard because it looked to some like 6s. The Fall 1991 issue of the *Skeptical Inquirer* notes that "the number 666 is linked to Satan in the Book of Revelation, and this helped fuel the false rumors fostered by fundamentalists." A federal judge in Topeka, Kansas, has approved settlements in the last of a dozen lawsuits filed by Procter & Gamble Co. to halt rumors associating the company with Satanism.
- President Ronald Reagan altered his California address to avoid the number 666. His name becomes 666 if you count the letters in his name.
- Why is it that when you add up the Roman numerals I = 1, V = 5, X = 10, L = 50, C = 100, and D = 500, you get 1 + 5 + 10 + 50 + 100 + 500 = 666?
- On May 1, 1991, the British vehicle licensing office stopped issuing license plates bearing the numbers "666." The Winter 1992 issue of the *Skeptical Inquirer* reports two reasons given for the decision: cars with 666 plates were involved in too many accidents, and there were "complaints from the public."
- Pope Innocent IV, whose Latin name corresponds to 666, was branded an anti-Christ by his opponents.

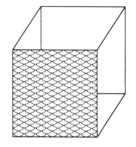

THE LOOM
OF GOD

The Pythagoreans associate six with marriage and health because six is the product of the first even and first odd numbers, which are female and male, respectively.

The number six also plays an important role in Islamic tradition. Islamic mystics consider our world a cubic cage, a six-sided prison. Humans struggle in vain to escape the bondage of the senses and physical world. Persian poets refer to such imprisonment as "six-door," or *shishdara*—the hopeless position of a gambler playing a form of backgammon. In this game, players use a six-sided die to move pieces. If an opposing player has locked (occupied) all six locations to which your playing piece could potentially have moved, then you are "shish-dar," or six-out, since your piece cannot move.

In Islam, the number 66 corresponds to the numerical value of the word *Allah*. Shown at left is an Islamic magic square that expresses the number 66 in every direction when the letters are converted to numbers. The square's grid is formed by the letters in the word *Allah*. Magic squares such as this were quite common in the Islamic tradition, but did not seem to have reached the West until the 15th century. From a historical perspective, my favorite Western magic square is Albrecht Dürer's which is drawn in the upper right hand column of his etching *Melencolia I* (see p. 77). The variety of small details in the etching have confounded scholars for centuries. Scholars believe that the etching shows the insufficiency of human knowledge in attaining heavenly wisdom, or in penetrating the secrets of nature.

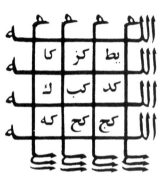

An Islamic magic square that expresses the number 66 in every direction. The grid is formed by the letters in the word *Allah*.

Dürer's 4×4 magic square, shown below, contains the first 16 numbers and has some fascinating properties. The two central numbers in the bottom row read 1514, the year Dürer made the etching. Also, in the vertical, horizontal, and two diagonal directions, the numbers sum to 34. In addition, 34 is the sum of the numbers of the corner squares (16 + 13 + 4 + 1) and of the small central square (10 + 11 + 6 + 7). The sum of the remaining numbers is 68 = 2 × 34. Was Dürer trying to tell us something profound about the number 34?

16	3	2	13
5	10	11	8
9	6	7	12
4	15	14	1

Mark Collins, a colleague from Madison, Wisconsin, with an interest in both number theory and Dürer's works, has studied the Dürer square and finds some astonishing features when converting the numbers to *hexadecimal binary code*. (In the binary representation, numbers are written in a positional number system that uses only two digits: 0 and 1.[1]) Since the first 16 numbers of this kind start with the number 0 and end with 15, he subtracts 1 from each entry in the magic square. Below is the result:

15	2	1	12
1111	0010	0001	1100
4	9	10	7
0100	1001	1010	0111
8	5	6	11
1000	0101	0110	1011
3	14	13	0
0011	1110	1101	0000

Melencolia I, by Albrecht Dürer (1514). This figure is usually considered the most complex of Dürer's works, the various symbolic nuances confounding scholars for centuries. Why do you think he placed a magic square in the upper right? Scholars believe that the etching shows the insufficiency of human knowledge in attaining heavenly wisdom, or in penetrating the secrets of nature.

A detail from the engraving *Melencolia I*.

Remarkably, if the binary representation for the magic square is rotated 45 degrees clockwise about its center so that the "15" is up and the "0" down, the resultant pattern has a vertical mirror plane down its center:

```
              1111
         0100      0010
     1000      1001      0001
 0011     0101      1010      1100
     1110      0110      0111
         1101      1011
              0000
```

For example, in row two, 0100 is the mirror of 0010. (I very much doubt that Dürer could have known about this symmetry.)

If we rotate the square counterclockwise so that the "12" is at the top and the "3" at the bottom, we get a pattern that has a peculiar left-right inverse when drawing an imaginary vertical mirror down the center of the pattern.

```
                1100
        0001          0111
  0010        1010          1011
1111    1001        0110          0000
  0100        0101          1101
        1000          1110
                0011
```

For example, in the second row, 0001 and 0111 are mirror inverses of each other.

Finally, Mark Collins has discovered the presence of mysterious intertwined hexagrams when the even and odd numbers are connected (see below).

I would be interested in hearing from those of you who find additional meaning in Dürer's magic square. Mark Collins and I are unaware of other magic squares having the symmetrical properties when converted to binary numbers. Mark has also done numerous experiments converting these numbers to colors, and comments:

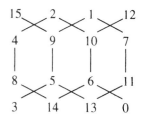

> "I believe this magic square is an archetype as rich in meaning and mysticism as the I-Ching. I believe it is a mathematical and visual representation of nature's origami—a photon of light."

Mark suggests you should create other mitosis-like diagrams by connecting 0 to 1 to 2 to 3. Then lift up your hand. Connect 4 to 5 to 6 to 7. Connect 8 to 9 to 10 to 11. Connect 12 to 13 to 14 to 15.

In concluding this section, I present a rather bizarre six-by-six magic square invented by the mysterious A. W. Johnson. All of its entries are prime numbers, and each row, column, diagonal, and broken diagonal sum to 666, the Number of the Beast. (By "broken diagonal," I mean the diagonal produced by wrapping from one side of the square to the other; for example, 131, 83, 199, 113, 13, 127 forms a broken diagonal).

The Apocalyptic Magic Square

3	107	5	131	109	311
7	331	193	11	83	41
103	53	71	89	151	199
113	61	97	197	167	31
367	13	173	59	17	37
73	101	127	179	139	47

The Seven Sleepers

The Pythagoreans call "7" a crisis, and consider all days of the month that could be divided by seven as critical days. The number seven is also important in Islamic tradition.

For example, the Koran states that God created Heaven and Earth in seven layers.[2] In addition, the seven Sleepers known from early Christian history also appear in the Koran (18:21). The names of the seven religious youths along with their dog are often presented in ornate Arabic calligraphic designs.

According to Christian legend, seven Christian boys of Ephesus hid themselves in a cave on Mount Coelian during the Decian persecution (A.D. 250). Eventually their hiding place was discovered, and its entrance blocked. The seven unfortunate youths fell asleep in mutual embrace. Then, about 200 years later, a herdsman rediscovered the cave and awoke the boys. One of the boys went to the market to buy food and was surprised to hear the name of Christ openly spoken. After the townspeople and the local emperor confirmed the presence of the boys in the cave, strengthening their faith in God and in the possibility of resurrection of dead, the boys fell asleep again.

An amulet from Turkey showing the names of the seven sleepers surrounding their special dog.

In another 9th century version of the legend, certain undecayed corpses of monks were found in a cave and thought to be the sleepers of Ephesus.

In the Koran (18:18), the seven Sleepers also have a dog with them in the cave:

> Thou wouldst have deemed them awake, whilst they were asleep, and we turned them on their right and on their left sides. Their dog stretched forth his two fore-legs on the threshold. If thou hadst come up on to them, thou wouldst have certainly turned back from them in flight, and wouldst certainly have been filled with terror of them.

There is an incredibly large number of occurrences of seven in all religions. Here are just a few examples:

- *Old Testament:* Lamech, father of Noah and son of the famous long-lived Methuselah, is born seven generations after Adam. Lamech lives for 777 years. (Did you think that 666 was the only important triplet for Christians?) Lamech should be avenged 77-fold (Genesis 4:24). Zechariah, a major biblical prophet, speaks of the seven eyes of the Lord. (Zechariah was concerned with the cosmic work of God, including the final judgment, restoration of Israel, and the coming of the messianic age.)

- *Sufism:* The idea of seven divine eyes occurs in Sufisim in connection with seven important saints who are the eyes of God. God is praised by creatures with 70,000 heads, each of which has 70,000 faces. There are seven points in the body upon which mystics concentrate their spiritual power.

- *Later Judaism:* Seven is important for Kabalists (see Chapter 19). In fact, Trachtenberg in his *Jewish Magic and Superstition* mentions the following cure for tertian (malarial) fever:

  > Take seven pickles from seven palmtrees, seven chips from seven beams, seven nails from seven bridges, seven ashes from seven ovens, seven scoops of earth from seven door sockets, seven pieces of pitch from seven ships, seven handfuls of cumin, and seven hairs from the beard of an old dog, and tie them to the neck-hole of the shirt with a white twisted cord.

- *Christianity:* The seven last words of Christ on the Cross. In the Book of Revelation, seven seals are opened, and there are seven trumpets blown to usher in the

Day of Judgment. Also: seven sacraments, seven gifts of the Holy spirit, seven deadly sins, seven layers of purgatory.

- *Islam:* The holy phrase: *la ilaha illa Allah Muhammad rasul Allah* has seven words ("There is no deity save God, Muhammad is the messenger of God.") Pilgrims must circle Kaaba in Mecca (Chapter 21) seven times.
- *Ancient Greece:* Swans circle seven times before Leto gives birth to Apollo. Apollo stays with the Hyperboreans for seven months.
- *Mithras:* In the ancient Irininan religion of Mithras, the souls rise through seven planetary spheres into the divine presence.
- *Vedas:* In India, Agni, the god of fire, has seven wives, mothers, or sisters as well as seven flames, beams, or tongues. The sun god owns seven horses.

Practicality of Number Theory

What could be more beautiful than a deep, satisfying relation between whole numbers. How high they rank, in the realms of pure thought and aesthetics, above their lesser brethren: the real and complex numbers ...
—Manfred Schroeder, *Number Theory in Science and Communication*, 1984

In this chapter, like others, we've discussed integer patterns both in geometry and mysticism. Is there anything special about integers in the fabric of the universe? Or do other kinds of numbers contribute equally to the loom upon which God weaves? Leopold Kronecker (1823–1891), a German algebraist and number theorist, once said, "The primary source of all mathematics are the integers." Since the time of Pythagoras, the role of integer ratios in musical scales has been widely appreciated. More importantly, integers have been crucial in the evolution of humanity's scientific understanding. For example, in the 18th century, French chemist Antoine Lavoisier discovered that chemical compounds are composed of fixed proportions of elements corresponding to the ratio of small integers. This was very strong evidence for the existence of atoms. In 1925, certain integer relations between the wavelengths of spectral lines emitted by excited atoms gave early clues to the structure of atoms. The near-integer ratios of atomic weights was evidence that the atomic nucleus is made up of an integer number of similar nucleons (protons and neutrons). The deviations from integer ratios led to the discovery of elemental isotopes (variants with nearly identical chemical behavior but with different radioactive properties). Small divergences in the atomic weight of pure isotopes from exact integers confirmed Einstein's famous equation $E = mc^2$ and also the possibility of atomic bombs. Integers are everywhere in atomic physics. Integer relations are fundamental strands in the mathematical weave— or as German mathematician Carl Friedrich Gauss said, "Mathematics is the queen of sciences—and number theory is the queen of mathematics."

CHAPTER **6**

St. Augustine Numbers

St. Augustine declared he had seen acephalic creatures, with eyes in their breasts....
> —*Glasgow Herald*, December, 31, 1924

Love with care—and then what you will, do.
> —St. Augustine

"Mr. Plex, I'm starting to worry." You gaze up at the statue of Aphrodite. Her left index finger is now several feet long and starting to puncture a nearby centaur's carotid artery. Strewn haphazardly beneath Aphrodite's feet are the discarded remains of coffee cups, french fry bags, and an occasional hamburger bun.

"Sir, I'll try to find a broom and sweep out our backroom."

"I'm not referring to the trash, you fool. It's the statues. I —I've a feeling our presence in the past is interfering with history."

"We're subtly changing the past?"

You nod. "Let's start packing our things this week." You motion to some notebook computers, the view-screen, a box of electronic flies....

Mr. Plex nods.

You sit down beneath Aphrodite. "Mr. Plex, while Pythagoras is cleaning his own house, I've something interesting to show you. Our agents discovered it in St. Augustine's time, A.D. 390, in Africa."

Mr. Plex comes closer and stares intently at you. You hand him a paper with the number "153" on it.

"St. Augustine, the famous Christian theologian, thought 153 was a mystical number, and that 153 saints would rise from the dead in the eschaton."

Mr. Plex rubs his forelimbs together. "How is that?"

"St. Augustine interpreted the Bible using numbers. For example, he was fascinated by a New Testament event (John 21:11) where the apostles caught 153 fish from the sea of Tiberias. Seven disciples hauled in the fish using nets. St. Augustine reasoned that these seven were saints."

"Sir, how could he know that?"

"Since there are seven gifts from the Holy Ghost which enable men to obey the ten commandments, he thought the disciples must therefore be saints. Moreover, $10 + 7 = 17$, and if we add together the numbers 1 through 17, we get a total of 153." You pause. "The hidden meaning of all this is that 153 saints will rise from the dead after the world has come to an end."

"It makes you wonder." Mr. Plex pauses and grins revealing his diamond teeth. "153 is my favorite integer."

"And why is that?"

"Because $153 = 1! + 2! + 3! + 4! + 5!$, where the '!' symbol means factorial. (Factorial is the product of all the positive integers from one to a given number. For example, $5! = 1 \times 2 \times 3 \times 4 \times 5 = 120$.)"

"Excellent, Mr. Plex, but better than that, $153 = 1^3 + 5^3 + 3^3$. Are there any other 3-digit numbers that equal the sum of the cubes of their own digits?"

"Mon Dieu! Here's something new." Mr. Plex pauses. "When the cubes of the digits of any 3-digit number which is a multiple of 3 are added, and the digits of the resulting number are cubed and added, and the process continued, the final result is 153. For instance, start with 369, and you get the sequence: 369, 972, 1080, 513, 153." Mr. Plex's limbs are shaking like a buzzsaw. "Could St. Augustine have known anything about this?"

"Curiouser and curiouser, Mr. Plex." You pause. "153 is also the 17th triangular number, and I just told you why St. Augustine thought 17 is important. It's the sum of 10 (for the ten commandments of the Old Testament) and 7 (for the gifts of the Spirit in the New Testament)."

"Back to triangular numbers again! Pythagoras would have liked all this. What could this all mean?"

"Just interesting coincidences, I'm sure."

Just then there is a knock on the door to your backroom hideout.

Mr. Plex goes over to the door. "Who's there?"

You try to pull him back from the door. "Shut up you fool."

A woman's voice comes from the other side of the door. "It's me, Theano."

Your heartbeat quickens as you walk over to the door and pull it open.

Theano hesitantly enters. From her ears dangle golden earrings in the shape of the tetraktys—the Pythagorean symbol consisting of 10 dots forming a triangle. Inscribed on her golden bracelets are various triangular and oblong numbers.

She holds herself like a queen. Her long neck curves like a bird taking wing. "I couldn't help coming back," she says while gazing into your eyes. "I want to learn."

At the husky tone of her voice, you shiver imperceptibly and all the air is expelled from your lungs in one wild gasp.

A fire within her seems to shoot upward and outward. "I want to learn."

The Loom of
God

St. Augustine Numbers

St. Augustine really was interested and excited about the number 153. St. Aurelius Augustinus Augustine was born in the year 354 and became the bishop of Hippo in Africa from A.D. 396–430. His mother, Monica, was Christian since girlhood. His father, Patricius, became Christian later in life.

St. Augustine's life was not always what normally associates with a saint. As a teenager, he became the unwed father of a boy he named Adeodatus. His mother was determined to find Augustine a proper marriage partner and get him married. She eventually found a "suitable" bride, who was not quite of marriageable age, and dismissed Adeodatus' mother, Augustine's true love. The whole incident distressed Augustine who proclaimed, "Give me chastity, but not yet!" Instead of marrying immediately, he found another girlfriend. (This chronology is not meant to disparage St. Augustine who was also a brilliant and spiritual man.)

Interestingly, St. Augustine's method of combining two influences ($10 + 7 = 17$) had its roots with the Pythagoreans. Recall how Pythagoras associated 5 with marriage because $5 = 2$ (female) $+ 3$ (male).

English mathematician Godfrey Hardy (1877–1947), in his book *A Mathematician's Apology*, gives $153 = 1^3 + 5^3 + 3^3$ as an example of a rare number which is the sum of the cubes of its digits. However he does not ascribe any mystical or mathematical profundity to this. In fact, he comments,

> These are odd facts, very suitable for puzzle columns and likely to amuse amateurs, but there is nothing in them that appeals to the mathematician. The proofs are neither difficult nor interesting—merely a little tiresome.

St. Augustine numbers, which are the sums of cubes of their digits, are a subset of a large class of numbers, which are the sums of powers of their digits. In other words, there are other n-digit numbers which are equal to the sum of the nth powers of their digits. Variously called narcissistic numbers, "numbers in love with themselves," Armstrong numbers, or perfect digital variants, these numbers have fascinated number theorists for decades. (The Appendix includes BASIC program code for computing St. Augustine numbers.)

The largest narcissistic number discovered to date is the incredible 39-digit number:

$$115132219018763992565095597973971522401$$

(Each digit is raised to the 39th power.) Can you beat the world record? What would Pythagoras have thought of this multidigit monstrosity?

As one searches for larger and larger narcissistic numbers, will they eventually run out? If they are proven to die out in one number system, does this mean they are finite in another? Martin Gardner wrote to me recently indicating that the number of narcissistic numbers has been proved finite. They can't have more than 58 digits in our standard base 10 number system.

Modern writer W. E. Bowman uses the number 153 throughout his book *The Ascent of Rum Doodle*. For example, there are 153 porters hired for a train which ascends the Himalayan mountains. The number 153 is also associated with the depth of a crevasse, the speed of the train, and the height of a ship above sea level.

I have not determined whether Bowman uses 153 because he is aware of St. Augustine's prediction regarding 153 saints rising from the dead when the world comes to an end.

Theano

There really was an important woman named Theano in ancient Greece. If you have not already guessed her identity, you will in the next chapter.

CHAPTER 7

Perfection

Six is a number perfect in itself, and not because God created all things in six days; rather the inverse is true; God created all things in six days because this number is perfect. And it would remain perfect even if the work of the six days did not exist.

—St. Augustine

Just as the beautiful and the excellent are rare and easily counted, but the ugly and the bad are prolific, so also abundant and deficient numbers are found to be very many and in disorder, their discovery being unsystematic. But the perfect are both easily counted and drawn up in a fitting order.

—Nichomachus, A.D. 100

Theano gazes up at the walls of your hideout which are now covered with mathematical symbols: oblong and triangular numbers, Doomsday equations, gnomons and St. Augustine numbers—all the result of your recent explorations with Mr. Plex.

You walk over to her. "Theano, we are travellers from the future studying the relationship between numbers, God, and the end of the universe."

You pause to assess the impact of your shocking revelation. When she just silently stares, you take her hand to comfort her.

Theano removes her hand from yours and wanders around the storage closet. It's a good thing she has not noticed the subtle transformations in the statues of the Greek gods.

She stops and looks at you with eyes as indecipherable as the air. "I'm impressed by your sophisticated devices."

Mr. Plex points with his insectile forearm. "Ma'am, they're called computers."

Why is Theano unafraid of you and Mr. Plex? Is it because the ancient Greeks were already familiar with strange ideas and gods, accustomed to a plethora of odd creatures: the river god Acheolous, the gorgon Medusa, Cerberus, winged lions, minotaurs, panther-birds, and harpys? Perhaps the transmogrification of the statues would also have little effect on her calm countenance.

You walk closer to her. "Theano, I'm impressed by your acceptance of all this."

Mr. Plex takes a few quiet steps toward you and whispers in your ear. "Shouldn't we be hiding all this modern technology from her? Think about what effect this might have on her ancient culture."

You push Mr. Plex away and turn to Theano. "Theano, soon we'll have to leave your time and return to the future. You weren't meant to see all this." You motion to the equations on the wall and the computers.

Her eyes are shrewd little chips of quartz. She grabs your hand. "Will you take me with you?"

"Impossible," Mr. Plex whispers.

You hesitate, but only for a second. "Yes, we'll take you."

Mr. Plex gasps and plants his forelimb over his mouth.

You continue. "But first we want to study Pythagoras a little longer."

You turn on the view-screen and see Pythagoras talking to Heraclitus. The sky is a perfect marine blue, the clouds rippling with iridescent orange and pink. Theano, Mr. Plex, and you settle down on the floor to watch.

Pythagoras raises his hand. "Heraclitus, I want to tell you about perfection." His voice is a whisper, as if he is afraid he is being watched.

"Perfection, Oh Great One?"

Pythagoras nods. "*Perfect numbers* are the sum of their divisors. For example, the first perfect number is 6 because $6 = 1 + 2 + 3$. The next perfect number is 28 because its divisors are 1, 2, 4, 7, and 14, and 28 also equals $1 + 2 + 4 + 7 + 14$."

Heraclitus' eyes seem to be locked onto Pythagoras' golden birthmark. "Pythagoras, there must be others."

"Yes, but we have not discovered them yet. These numbers are so rare that they must have a special significance for the Creator. He must have had a definite plan for them." Pythagoras pauses. "I think perfection is rare in numbers just as goodness and beauty are rare in humans. On the other hand, imperfect numbers are common, and so are ugliness and evil."

"Imperfect numbers?"

"Those where the sum of the factors is greater or less than the number itself. Imperfect numbers are like babies born with birth defects. The numbers have too many or too few fingers or organs."

Heraclitus nods. "My friend Milo told me about *abundant numbers*. Can you explain them to me?"

Pythagoras pinches his lower lip with his teeth. "How dare he reveal that secret." He then takes a deep breath. "If the original number is less than the sum of its factors, I call it *abundant*. As an example, the factors of 12 are 1, 2, 3, 4, and 6. And these factors add up to 16. If greater, the number is *deficient*. For example, the factors of 8—1, 2, and 4—add up only to 7."

"Most numbers are either abundant and deficient? Perfection is rare."

Pythagoras nods. "You've got it!" Then he leans toward Heraclitus as if observing a painting in a museum. "Heraclitus, two numbers are *amicable* or friendly if the sum of the divisors of the first number is equal to the second number, and vice versa. They have the same parentage and in their divine world are more congenial than numbers which are unfriendly."

"I don't get it."

"Here's an example. 220 and 284 are amicable. Let's list all the numbers by which 220 is evenly divisible."

Heraclitus leans forward and clasps his hands together like an eager child. "Uh, let's see—1, 2, 4, 5, 10, 11, 20, 22, 44, 55, and 110 all go into 220."

"Excellent, now add up all those divisors. What do you get?"

"$1 + 2 + 4 + 5 + 10 + 11 + 20 + 22 + 44 + 55 = 284$."

"Very good, Heraclitus. The answer is 284. Now let's try the same trick with 284. Its perfect divisors are 1, 2, 4, 71, and 142. Now, add them up."

"You get 220, Oh Great One."

"Yes! Therefore 220 and 284 are amicable numbers. The sums of their divisors are equal to each other."

Heraclitus nods. "Interesting. Amicable numbers, like perfect ones, are quite rare."

"War is always easier than peace."

"220 and 284 would be perfect marriage partners in the eyes of God."

Pythagoras nods. "A perfect marriage."

数 数 数 数

Theano clenches her fists as you go over to the view-screen and stab your finger at the on/off button. Pythagoras and Heraclitus fade like ghosts. She then spins around and faces you. "He cares a bit too much about numbers." She clenches her fists. "Sometimes, I wonder if I made the right choice to marry him."

Your mouth hangs open. "You–you mean you're Pythagoras' wife?"

Mr. Plex taps his head with his forelimb. "Sir, you are dense."

You swivel around, shocked by the tone of Mr. Plex's voice. Then you turn back to Theano.

Theano rubs her hands together. "It's not that I'm uninterested in mathematics—"

You shrug and take a deep breath. "There's so much to study," you say, "so much to understand. Mathematics is the key to enlightenment."

"True," she says with a trace of uncertainty.

You scan Theano's eyes for a glimmer of understanding. A little fire dances in your own eyes.

"Yes, mathematics is the key," she says. "It's so pure. Infinite."

"What's wrong?"

"By Zeus, I'm not sure."

You feel like shaking her. "Theano, stop thinking about your gods. Stop relying on invisible gods on Mount Olympus to give your life meaning. There is no Zeus, no Apollo, no Aphrodite. Think for yourself."

"I am. I am." She comes closer and gives your hand a squeeze. "Don't be angry. I agree mathematics is beauty. I don't know why I'm scared. Please be patient." She pauses. "This—this is all so over my head."

"Theano, I'll teach you. Guide you. I'll always be with you."

Mr. Plex gasps. "Sir, the *Main Directive*—"

You look into Theano's dark eyes. "You're having trouble imagining all of this. I sympathize." You give her a quick hug.

Mr. Plex makes a gagging sound. Perhaps he is making fun of your sentimental exchange with Theano.

You regain your composure and gaze from Theano to Mr. Plex. "Let me tell you both more about perfect numbers."

Mr. Plex withdraws a notebook computer from the abdomen of Aphrodite who has now sprouted a third arm. Evidently, your presence in the past continues to spawn a parallel universe subtly different from your own. Mr. Plex looks at you. "There's no need to keep this hiding place secret any longer," he says. He remains poised and ready to type.

You go to a wall of the temple closet and scrawl on it with a piece of charcoal. You lower your voice an octave, and you think you see awe in Theano's eyes. "The first four perfect numbers, 6, 28, 496 and 8128 were known to the late Greeks. Nicomachus and Iamblichus—Pythagoreans who were born after your husband died —knew about these."

Theano raises her hand. "Do all perfect numbers end in an 8 or 6?"

"I'm not sure. But I do know that every even perfect number is also a triangular number." You pause. "Perfect numbers are very rare. The fifth perfect number 33,550,336 was found recorded in a medieval manuscript. To date, mathematicians know only about 30 perfect numbers. No one knows if the number of perfect numbers is infinite."

As before, a chill goes up your spine when you say the word "infinite."

You begin to pace. "Perfect numbers thin out very quickly as you search larger and larger numbers. They might disappear completely—or they might continue to hide among the multidigit monstrosities that even our computers can't find."

Mr. Plex raises his forelimb. "What about amicable numbers?"

You nod. "Over a thousand amicable numbers have been found. Another pair includes 17,296 and 18,416."

Mr. Plex begins to furiously type a program to search for and print amicable numbers. The computer soon prints several numbers on a slip of paper:

```
Amicable Numbers
  220    284
 1184   1210
 2620   2924
 5020   5564
 6232   6368
10744  10856
```

"Good work, Mr. Plex."

Theano takes the slip of paper and studies it.

You continue your discussion. "Mathematicians have even studied *sociable* numbers. In these sets of number, the sum of the divisors of each number is the next number of a chain. For example, our agents in 1918 spied on a man named Poulet who found the following sociable number chain:

$$12,496 \rightarrow 14,288 \rightarrow 15,472 \rightarrow 14,536 \rightarrow 14,264 \rightarrow 12,496$$

Sociable chains always return to the starting number. Poulet's chain, and a 28-link chain starting 14,316, were the only sociable chains known until 1969 when suddenly Henri Cohen discovered seven new chains, each with four links."

Your voice grows in intensity and speed. "A pair of amicable numbers, such as 220 and 284, is simply a chain with only two links. A perfect number is a chain with only one link." You take a deep breath. "No chains with just three links have been found, despite massive searches. There are certainly none with a smallest member less than 50 million! These hypothetical 3-link chains are called *crowds*. Mathematically speaking, a crowd is a very elusive thing, and may not exist at all."

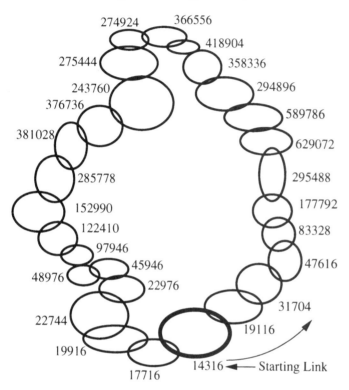

A wonderful 28-link amicable number chain.

"Sir, you talk about discovering numbers as if we're searching for stars in the heavens."

"It's a little like that. There's a lot of unexplored territory. Wait until I tell you about fractals …"

There is a crackling sound as Theano unwraps some food from its paper wrapping. "What do you call these things. They're quite good." She wipes a fleck of saliva from her lips.

You bend down for a closer look. "Hamburgers," you say.

"From McDonald's," Mr. Plex says.

Just then the floor begins to shake. Theano and Mr. Plex look warily from one to the other like condemned criminals.

"Sir, we're destroying this world. I don't know why, but perhaps the act of going back in time creates all sorts of problems and paradoxes, and tears the spacetime fabric. We never should've stayed here so long. What if the Pythagoreans found one of our computers and it changed their religion? The future history of their world would be irrecoverably altered."

You take a few steps toward Theano and notice something peculiar about the equations on the wall's surface behind her. What!? With a quick intake of breath like someone about to plunge into icy water, you walk even closer.

Upon the wall are hundreds of small, moist creatures resembling shrimp. Their exoskeletons glow an iridescent gold.

For a moment, the crustaceans and the translucent statues of the Greek gods appear to be coming toward you. You feel trapped within a tangle of golden reflections as demented shrieks fill the Temple of Apollo like the prayers of the dying.

Aphrodite's head is the color of ice, surrounded by tiny shrimp phantoms that scream and wriggle. The statues of the gods run yelling through the temple's backroom, a mineralized mirage, away from the world of man.

Or so it seems. You shake your head, and the statues appear never to have moved.

Nothing moves. A coldness seems to rise from the marble beneath the three of you. Tiny crustaceans congregate in the shadows. You watch them crawl and grow still.

A polyp has sprouted from Aphrodite's smooth neck. It looks lurid. You have the urge to pull it out. But what is happening to it? You stare, the cold heavy on your eyelids, the crustaceans rising. Was the polyp actually growing?

No. Yet how enormous it looks. The cold, the stillness, the sudden coming of the gods like intruders into your domain at the moment you were connecting with Theano. You can be sure of nothing.

You take a napkin from a pile and rub the cold sweat from your forehead. You walk closer to Theano. You are confused, unsure, uncertain that anything unusual has happened.

All of your plans for teaching more mathematics to Theano and Mr. Plex come back to you. So much to do.

The tremors in the Earth start again causing a huge centaur statue to begin to crash down upon you. "Mr. Plex! Theano!"

Mr. Plex leaps toward you and covers your body just as the statue is about to slam down upon you with all its mass. There is a cracking sound, and you don't know if it is the statue breaking or Mr. Plex's back.

THE SCIENCE BEHIND THE SCIENCE FICTION

> *Pythagoras was intellectually one of the most important men that ever lived, both when he was wise and when he was unwise.*
> —Bertrand Russell, *A History of Western Philosophy*

Amicability

Pythagoras' description of perfect, imperfect, and amicable numbers is based on historical records. For example, he felt these numbers were so rare that they had a special significance for the Creator. He also felt that perfection was rare in numbers just as goodness and beauty were rare in humans.

We know from Iamblichus that Pythagoras really did list the pair of numbers 220 and 284 as an example of friendly or amicable numbers. Iamblichus refers to amicable pairs as friends or "alter egos." Numbers are seen as divine archetypes—in God's mind from the beginning. As a result, Pythagoreans believed that the study of numbers is literally a way to decipher the divine plan.

In terms of human amicability, when you and Mr. Plex meet Theano, she appears to be torn between her interest in spending time with you and her loyalty to Pythagoras. Her uncertainty is understandable, because all historical evidence suggests Pythagoras to be a model husband and father, although it seems at times he had a touch of arrogance. He attributed to himself a semidivine character, and appears to have said, "There are men and gods, and beings like Pythagoras." These feelings of self-supremacy could conceivably place stress on any marriage.[1]

Pythagoras is one of the most interesting and puzzling men in history. He founded a religion, of which the main tenets were transmigration of the souls and the sinfulness of eating beans. But the unregenerate hankered after beans, and sooner or later rebelled.
—Bertrand Russell, *A History of Western Philosophy*

In Chapter 3, I mentioned some of the Pythagorean taboos. Here are some additional ones:

1 Do not eat a heart.
2 Do not step over a crossbow.
3 Do not walk on highways.
4 Do not look in a mirror next to a light.
5 Do not leave a mark in ashes when taking a pot off a fire. (Instead, stir the ashes to erase the mark.)
6 Do not allow birds to inhabit the roof of your house.
7 When you rise from the bedclothes, roll them together and smooth out the impression of the body.

Perfect Numbers

Man ever seeks perfection but inevitably it eludes him. He has sought "perfect numbers" through the ages and has found only a very few—twenty-three up to 1964.
—Albert H. Beiler, *Recreations in the Theory of Numbers*

As you told Mr. Plex, the first four perfect numbers, 6, 28, 496, and 8128 were known to the late Greeks and Nicomachus, a disciple of Pythagoras. Perfect numbers are indeed difficult to find.

The first ten perfect numbers are:

1   $2 M_2 = 6$
2   $2^2 M_3 = 28$
3   $2^2 M_5 = 496$
4   $2^6 M_7 = 8128$
5   $2^{12} M_{13} = 33550336$
6   $2^{16} M_{17} = 8589869056$ (Discovered in 1588 by Cataldi)
7   $2^{18} M_{19} = 137438691328$ (Discovered in 1588 by Cataldi)
8   $2^{30} M_{31} = 2305843008139952128$ (Discovered in 1772 by Euler)
9   $2^{60} M_{61}$ (Discovered in 1883 by Pervusin)
10 $2^{88} M_{89}$ (Discovered in 1911 by Powers)

The 30th perfect number, $2^{216090} M_{216091}$ was found using a CRAY supercomputer in 1985 (see Table 7.1).

To understand this list of the first ten perfect numbers, first note that perfect numbers can be expressed as: $2^X(2^{X+1} - 1)$ for special values of X. Euclid proved that this rule was sufficient for producing a perfect number, and Euler, 2000 years later, proved that all even perfect numbers have this form, if $2^N - 1$ is a prime number. Such numbers are called

Table 7.1.   All Known Perfect Numbers

93

Perfection

$2^{N-1} \times (2^N - 1)$ Number	N	Discovered (year, human)
1–4	2, 3, 5, 7	in or before the middle ages
5	13	in or before 1461
6–7	17, 19	1588 Cataldi
8	31	1740 Euler
9	61	1883 Pervouchine
10	89	1911 Powers
11	107	1914 Powers
12	127	1876 Lucas
13–17	521, 607, 1279, 2203, 2281	1952 Robinson
18	3217	1957 Riesel
19–20	4253, 4423	1961 Hurwitz & Selfridge
21–23	9689, 9941, 11213	1963 Gillies
24	19937	1971 Tuckerman
25	21701	1978 Noll & Nickel
26	23209	1979 Noll
27	44497	1979 Slowinski & Nelson
28	86243	1982 Slowinski
29	110503	1988 Colquitt & Welsh
30	132049	1983 Slowinski
31	216091	1985 Slowinski
32?	756839	1992 Slowinski & Gage
33?	859433	1993 Slowinski

Mersenne prime numbers M_N after their inventor Marin Mersenne. For example, 127 = $2^7 - 1$ is the 7th Mersenne number, denoted by M_7, and it is also prime and the source of the fourth perfect number, $2^6 M_7$. (Mersenne prime numbers are a special subclass of Mersenne numbers generated by $2^N - 1$.)

As with many of the best mathematicians, father Marin Mersenne was a theologian. In addition, Father Mersenne was a philosopher, music theorist, and mathematician. He was friend of Descartes, with whom he studied at Jesuit college. Mersenne found several prime numbers of the form $2^N - 1$, but he underestimated the future of computing power by stating that all eternity would not be sufficient to decide if a 15- or 20-digit number were prime. Unfortunately the prime number values for N which make $2^N - 1$ a prime number form no regular sequence. For example, the number is prime when $N = 2,3,5,7,13,17,19, \ldots$ Notice when N is equal to the prime number 11, $M_{11} = 2047$ which is not prime because $2047 = 23 \times 89$.

In 1814, P. Barlow in *A New Mathematical and Philosophical Dictionary* wrote that the 8th perfect number was "probably the greatest perfect number that ever will be discovered for they are merely curious without being useful, and it is not likely that any person will attempt to find one beyond it." Barlow placed such a limit on human knowledge because, before computers, the discovery of Mersenne primes depended on laborious human computations. M_{31} or $2^{31} - 1 = 2147483647$ is quite large, even though Euler in 1772 was able to ascertain that it is a prime number.

With the electronic computer, Barlow's limit on humanity's knowledge was rendered invalid. Because of their special form, Mersenne numbers are easier to test for primality than other numbers, and therefore all the recent record-breaking primes have been Mersenne numbers—and have automatically led to a new perfect number.

There is a bizarre and puzzling relationship between cubes and perfect numbers. Every even perfect number, except 6, is the sum of the cubes for consecutive odd numbers. For example,

$$28 = 1^3 + 3^3$$
$$496 = 1^3 + 3^3 + 5^3 + 7^3$$
$$8128 = 1^3 + 3^3 + 5^3 + 7^3 + 9^3 + 11^3 + 13^3 + 15^3$$

(The Appendix contains a BASIC program listing for computing perfect numbers.)

Odd Perfect Numbers

Odd perfect numbers are even more fascinating than even ones for the sole reason that no-one knows if odd perfect numbers exist. They may remain forever shrouded in mystery. On the other hand, mathematicians have catalogued a long list of what we *do* know about odd perfect numbers; for example, we believe that none will be discovered with values less than 10^{200}. Mathematician Albert H. Beiler says, "If an odd perfect number is ever found, it will have to have met more stringent qualifications than exist in a legal contract, and some almost as confusing." Here are just a few.

An odd perfect number

1   must leave a remainder of 1 when divided by 12 or a remainder of 9 when divided by 36.
2   must have at least six different prime divisors.
3   if not divisible by 3, must have at least 9 different prime divisors.
4   if less than 10^{9118}, is divisible by the 6th power of some prime.

Author and mathematician David Wells comments, "Researchers, without having produced any odd perfects, have discovered a great deal about them, if it makes sense to say that you know a great deal about something that may not exist." Similarly, Sufi scholar Clark Punkord remarks, "Theologians, without proof of God, have defined many characteristics of God, but can one meaningfully catalog characteristics about something that may not exist?"

Judaism and Perfect Numbers

Although the Jews of the Hellenistic world made fun of the ancient Greek legends, they gradually found themselves very fond of using numbers for validating the scriptures.

The Jews gradually realized that many legends and numbers of the Old Testament had Greek counterparts.

The ancient Jews tried to use numbers to prove to atheists that the Old Testament was part of God's revelation. It was therefore fortuitous that the Old Testament was chock full of numerical references for the Jews to interpret. For example, Philo justified the story of Genesis by first validating the assertion that God created the world in six days. He claimed that the six-day creation must be correct because 6 was the first perfect number (since 6 is the sum of its divisors, $6 = 1 + 2 + 3$). The perfection of the number 6 led to it becoming the symbol of creation and reinforcing the notion of God's existence. Despite the fact that the Jews scoffed at the Greek legends and numerology, they began to use a Pythagorean concept to reinforce the notion that Jahweh was the one true God.

So important were perfect numbers to the Jews in their search for God that Rabbi Josef ben Jehuda Ankin, in the 12th century, recommended the study of perfect numbers in his book *Healing of the Souls*.

Why did the ancients have such a fascination with numbers? Could it be that in difficult times numbers were the only constant thing in an ever-shifting world? Numbers were tangible, immutable, comfortable, eternal, more reliable then friends, less threatening than Zeus.

Christianity, Islam, and Perfect Numbers

Christians in the 12th century were very interested in the second perfect number 28. For example, since the lunar cycle is 28 days, and because 28 is perfect, philosopher and theologian Albertus Magnus[2] (A.D. 1200–1280) expressed the idea that the mystical body of Christ in the Eucharist appears in 28 phases.

The perfect number 28 also plays an important role in Islam, for religious Moslems connect the 28 letters of the alphabet in which the Koran is written with the 28 "lunar mansions." For example, the famous medieval mathematician al-Biruni (1048) suggested that this connection proves the close relation between the cosmos and the word of God. Note also that the Koran names 28 prophets before Mohammed.

Peter Bungus, Numerorum Mysteria, and Dr. Krieger

Throughout both ancient and modern history, the feverish hunt for perfect numbers became a religion. The mystical significance of perfect numbers reached a feverish peak around the 17th century. Peter Bungus, for example, was one of a growing number of 17th century mathematicians who combined numbers and religion. In his alchemic book titled *Numerorum Mysteria*, he listed 24 numbers said to be perfect, of which Mersenne later stated that only 8 were correct. Mersenne went on to add three more perfect numbers, for $N = 67$, 127, and 257 in an equation which can be used for *even* perfect numbers $(2^{N-1})(2^N - 1)$, but it took a walloping 303 years before mathematicians could check Mersenne's statement to find errors in it. 67 and 257 should not be admitted, and perfect numbers corresponding to $N = 89$ and 107, for which the Mersenne numbers are prime, should be added to the list.

How could Mersenne, back in the 17th century, have conjectured about the existence of such large perfect numbers? After centuries of debate, no one has an answer. Could he have discovered some theorem not yet rediscovered? Recall that empirical methods of his time could hardly have been used to compute these large numbers. (The Mersenne number for $N = 257$ has 78 digits.)

Zealous attempts at perfection are not limited to Peter Bungus and Mersenne. Even in the 1900s there have been startling attempts to find the holy grail of huge perfect numbers. For example, on March 27, 1936, newspapers around the world trumpeted Dr. S. I. Krieger's discovery of a 155-digit perfect number ($2^{256}(2^{257} - 1)$). He thought he had proved that $2^{257} - 1$ is prime. The Associated Press release, appearing in the *New York Herald Tribune* (March 27, 1936, 95: 21), was as follows:

PERFECTION IS CLAIMED FOR 155-DIGIT NUMBER
Man Labors 5 Years to Prove Problem Dating from Euclid

Chicago, March 26 (AP)—Dr. Samule I. Krieger laid down his pencil and paper today and asserted he has solved a problem that had baffled mathematicians since Euclid's day—finding a perfect number of more than 19 digits.

A perfect number is one that is equal to the sum of its divisors, he explained. For example, 28 is the sum of 1, 2, 4, 7, and 14, all of which may be divided into it. Dr. Krieger's perfect number contains 155 digits. Here it is: 26, 815, 615, 859, 885, 194, 199, 148, 049, 996, 411, 692, 254, 958, 731, 641, 184, 786, 755, 447, 122, 887, 443, 528, 060, 146, 978, 161, 514, 511, 280, 138, 383, 284, 395, 055, 028, 465, 118, 831, 722, 842, 125, 059, 853, 682, 308, 859, 384, 882, 528, 256.

Its formula is 2 to the 513th power minus 2 to the 256th power. The doctor said it took him 17 hours to work it out and five years to prove it correct.

Unfortunately for Dr. Krieger, a few years earlier the number $2^{257} - 1$ was found to be composite (non-prime). Mathematical journals during the time of Krieger therefore wrote letters to the *New York Herald Tribune* complaining that it had sacrificed accuracy for sensationalism.

The Book of Genesis

Like perfect numbers, amicable numbers have also fascinated religious and mystically inclined individuals through the ages. For example, in Genesis 32:14, Jacob gives a present of 220 goats:

And he lodged there that night; and took of that which he had with him a present for Esau his brother: two hundred she-goats and twenty he-goats, two hundred ewes and twenty rams....

According to several theologians, this was a "hidden secret arrangement" because 220 is one of a pair of amicable numbers 220-284, and Jacob tried this means of securing friendship with Esau. Pythagoras himself said that a friend was "one who is the other 'I' such as are 220 and 284."

El Madshriti, an Arab of the 11th century, experimented with the erotic effects of amicable numbers by giving a beautiful woman the smaller number 220 to eat in the form of a cookie, and eating the larger 284 himself! I am not sure whether his mathematical approach to winning the woman's heart was successful, but this method may be of interest to all modern dating services. Imagine restaurants of the future branding the numbers into two pieces of filet mignon for two prospective marriage candidates. Perhaps amicable number tattoos will one day be used for mathematical displays of public affection.

Our Arab friend, El Madshriti was not the last to make use of amicable numbers to unite the sexes. In the 14th century, the Arab scholar Ibn Khaldun said in reference to amicable numbers:

> Persons who occupy themselves with talismans assure that these numbers have a particular influence in establishing union and friendship between two individuals. One prepares a horoscope theme for each individual. On each, one inscribes one of the numbers just indicated, but gives the *strongest number* to the person whose friendship one wishes to gain. There results a bond so close between the two persons that they cannot be separated.

When Ibn Khaldun used the term "strongest number," he was not certain whether to use the larger of the two amicable numbers or the one that had the most divisors.

Since antiquity, Arabs have been interested in different ways for finding amicable numbers. One personal favorite is taken from the Arabian mathematician/astronomer Thabet ben Korrah (A.D. 950). Select any power of 2, such as 2^x and form the numbers

$$a = 3 \times 2^x - 1$$
$$b = 3 \times 2^{x-1} - 1$$
$$c = 9 \times 2^{2x-1} - 1$$

If these are all primes, then $2^x ab$ and $2^x c$ are amicable. When x is 2, this gives the numbers 220 and 284. (The Appendix contains a BASIC program listing for computing amicable numbers.)

The Story of the 672nd Night

The number 672 is one of many *multiply perfect numbers*—numbers such that the sum of *all* their divisors is an exact *multiple* of the number. For example, 120 is a triple perfect number because its divisors $1 + 2 + 3 + 4 + 5 + 6 + 8 + 10 + 12 + 15 + 20 + 24 + 30 + 40 + 60 + 120$ add up to 360 which is 3×120. Similarly, 672 is a triple perfect number.

There have been several recent attempts to explain the mysterious title of Hugo von Hoffmannsthal's tale *The Story of the 672nd Night*. Hugo von Hoffmannsthal (1874–1929), was an Austrian poet, dramatist and essayist, best known for writing libretti for Richard Strauss' operas. One explanation for his title is the fact that 672 is a multiply perfect number, but no literary scholars are certain if this is Hoffmannsthal's reason for using 672.

Some scholars suspect that the 672 in *The Story of the 672nd Night* is connected with the tale *1001 Arabian Nights*. Incidentally, the tale was originally titled "Das Märchen der 672 Nacht," and first published in 1895.

A Behemoth Perfect Number

As already mentioned, an even perfect number has the form $2^{N-1} \times (2^N - 1)$. Harry J. Smith of Saratoga, California wrote a program using Borland C++ to compute a perfect number if given the exponent of a Mersenne prime. For the largest known perfect number ($N = 859433$, see Table 7.1), his result is an output file 530,462 bytes long:

```
8.38488226750157042500640354808866515253415429199140040599280285821709907899741388130216829400311620770395775546150046571822455435668593586828368892640169238254781909679756198109742366665971528473510148532200207765550761429250223709343984391870127237160530335863712225120844478400702832379424390413335487569466042025205673357930282035736053809815194204427457703738579711501797104277768652841386730184774423285023243335004300497700439435717417921596566014755539083543277350197922333... 79824650626204001549281235248638211738314872927426303409137837262613583698080504676779300361990978586055050893978728388987879499444003227699065497769734977285013205018849822209520636324006280946502690084068393342022747979424136073779133409413621366454724565364104439784424412588149743051467775011265 1540416167936 E+517429
```

The " ... " indicates where digits have been deleted to shorten this chapter!

The Death of Perfection

No one knows if perfect numbers eventually die out as one sifts through the landscape of numbers. The mathematical landscape is out there, waiting to be searched. The Pythagoreans could only find four perfect numbers, and we can find over 30. Will humanity ever discover more than 40 perfect numbers? There is a limit on humans' mathematical knowledge arising not only from our limited brains but also our limited computers. In a strange way, the "total" of mathematical knowledge is godlike—unknowable and infinite. As we gain more mathematical knowledge, we grow closer to this god, but can never truly reach him. All around us we catch glimpses of a hidden harmony in the works of humans and nature. From the great pyramid of Cheops to patterns in plants, we see evidence of design by precise geometrical laws. Nobly, we continue to search for the connections underlying all that is beautiful and functional.

CHAPTER 8

Turks and Christians

I do not know what, if anything, the Universe has in its mind, but I am quite, quite sure that, whatever it has in its mind, it is not at all like what we have in ours. And, considering what most of us have in ours, it is just as well.
—Ralph Estling, *The Skeptical Inquirer*, 1993

Cannot there be found a Christian to cut off my head?
—Emperor Constantine XI Palaeologus, May 29, 1453

"I don't like this one bit," Mr. Plex says as he gazes up at some Greek statues and dusts himself off.

Hercules' esophagus is evaginating through his mouth, and Aphrodite's head looks like a moose.

You gaze down at the remains of the large centaur statue that had crashed down upon Mr. Plex's back. "Thank you, Mr. Plex, for saving my life."

"It was nothing, sir. My diamond body can withstand a thousand times that weight."

Theano shudders. "What's happening?" she says. "We've had minor earthquakes around here, but nothing like that."

You walk closer to her. "Time travel is always risky. Especially when the transfinites are watching."

"Transfinites?"

You nod. "They're angry with us and causing all this mischief with the statues." You pause when you see a confused expression on the faces of Mr. Plex and Theano. "They're a race of advanced beings who don't like us traveling back to the past. I guess they want to scare us. The further back we travel in time, the more angry they seem to get."

"What about the crustaceans, sir?"

"They're the 'keepers of time.' Whenever the transfinites break through to try to scare us back to the future, the shrimp try to repair the crack in time. But there's no need to worry. We'll have another week before we really have to get the hell out of here."

毃

Mr. Plex is breathing rapidly. "Sir, where do transfinites come from?"

"The Tarantula Nebula. It's a mix of gas and dust in the Large Magellanic Cloud, one of the Milky Way's two companion galaxies."

"Far away?"

You nod. "Although it's 190,000 light-years away, you could see it with your naked eye in the Earth's southern hemisphere." You pause. "Somehow they've learned to use a wormhole in space to quickly travel the vast distances."

Theano stretches her arms and begins to pace. "Can we get out of this room?"

You kick at some discarded hamburger wrappers. "Why not?"

"Sir—"

"It's OK, Mr. Plex. We'll just walk around *inside* the Temple."

"But sir, we might be seen between the columns."

"Relax, Mr. Plex."

It is early morning. Somewhere in the distance, a lone kithara player strums a ragged, soulful song. As you gaze out from between the massive columns, the sky seems limitless, gleaming with a magical light beyond an endless terrain of hills. The scent of rain mingles with the cool breeze.

"I love the fresh air," Theano says.

She begins to talk now and then in easy bursts, her smooth voice always charming and distracting you slightly from the content of what she says. Now Theano seems eager, happy, curious about things around her. No regrets apparently. But then it is so soon …

You lean upon a massive marble column and begin to draw on its surface:

X X X X X . . . . X X . . X . X X . X . X . . X . . X X . .

"Sir, what is it? Some kind of code?"

"Let me tell you what our agents in the year 1453 have just discovered—diabolical mathematics involving two religions. In the 1950s the scenario evolved into a game called *Turks and Christians*.

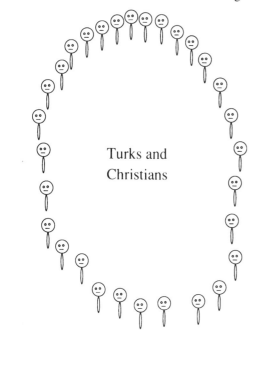

Turks and Christians

Theano drifts closer. "Sounds fascinating," she says, "but give me some background. What's this about the year 1453?"

You nod. "The mathematical puzzle involves the concept of *decimation*. Throughout history, the lawful penalty for mutiny on a ship was to execute one-tenth the crew. First the crew members were forced to randomly line up in a circle. It was then customary that the victims be selected by counting every 10th person from the circle, hence the term *deci*mation. Eventually this term has come to be applied to any depletion of a group by any fixed interval—not just ten."

Theano appears deep in thought. Her golden tetraktys earrings glisten in the sunlight. "What does this have to do with religion?"

"A very old puzzle about decimation is called *Turks and Christians*. The story is that 15 Turks and 15 Christians were aboard a sailing vessel caught in a hurricane. The captain decided to appease the gods by throwing half his passengers overboard—"

"Horrible story, sir."

"In order to leave the selection of victims to chance, the captain arranged all 30 people in a circle, and he announced to them that every thirteenth person would be killed. A clever but ruthless Christian whispered to his fellow Christians a plan on how they might be saved. He pointed out exactly how to take places in the circle so that only 15 Turks would be counted and thrown overboard."

"Sir, is that possible?"

Theano whispers, "Why did the Christians hate the Turks?"

"Perhaps because the Turks killed the Christian emperor Constantine XI Palaeologus in the year 1453 and slaughtered the Christians in Constantinople."

Theano and Mr. Plex are deep in thought as they try to determine the arrangement of 15 Turks and 15 Christians that permits the Christians to live but sets up the Turks for certain death.

You begin to draw on the column. "Here, let me give the answer. This diagram shows the arrangements of Christians and Turks. I've represented the circle of men by a line. The last Turk stands next to the first Christian in the circle."

```
C C C C C  T  T T T C C T T C  T  C C  T  C  T  C T T C T T C C  T   T
X X X X X 12 5 7 3 X X 9 1 X 15 X X 11 X 14 X 6 4 X 8 2 X X 13 10
```

"If you start counting at the first man, you'll see that the Turks will be counted out first, in the order indicated by the numbers. For example, the 1st Turk thrown over the ship is marked as a "1". The 15 X's are the Christians who survive.

Theano studies your diagram. "The last poor soul is marked 15," she says. "He'll see his screaming comrades all thrown overboard and probably never get the chance to realize the Christians orchestrated the whole thing."

Mr. Plex pulls a notebook computer from beneath a huge deformed statue of Apollo and types feverishly. "Sir, it seems this puzzle can be solved purely mechanically. I can have 30 array elements in my program, mark my starting point, decimate by 13, and note which are the first 15 elements to be counted out."

"Quite so."

Far away, over the hills, among the camellias, you think you see someone moving. You hear steps echoing. But it's only the rustling of leaves.

You turn back to Mr. Plex and Theano. "Here's another decimation puzzle. Fiendishly

Turks being thrown over the side of a ship.

difficult. Five mortals and five gods found five emeralds. They soon began to fight as to who should get the jewels. Zeus bellowed, 'We'll arrange ourselves in a circle, and count out individuals by a fixed interval, and give an emerald to an individual as he leaves the circle.' Zeus was clever and arranged the circle so that by counting a certain mortal as 'one' he could count out all the gods first. The arrangement of gods and mortals was:

1 2 3 4 5 6 7 8 9 10
M M M G G G M G G   M

The count starts with the mortal at the extreme left, goes to the right, and then returns to the leftmost individual remaining. Each individual 'counted out' steps out of the circle and is not included in the count thereafter.''

You pause, and grin. "Plato was the mortal who was counted as one, and he insisted on his right to choose the interval of decimation. Again, the count starts with Plato, the mortal at the extreme left. Plato's astute choice counted out all the mortals first. What interval will count out the five gods first, and what will count out the five mortals first?''

Theano and Mr. Plex study the problem. After several minutes, Theano lets out a whistle of surprise and gives you the two solutions.

"Mon Dieu!" screams Mr. Plex.

You take a quick breath of utter astonishment. "Theano, you're a genius!" You run over and shake her hand.

Theano draws her lips into a tight smile. "Thank you.''

Before you have a chance to congratulate her further, you hear running sounds coming from the valley near the temple.

Theano puts her hands over her forehead to shield the sun. "Oh, no,'' she yells as she points. "It's my husband.''

You squint and see Pythagoras running toward you swinging his kithara like a club. "Sir—''

"Not now Mr. Plex.''

"Sir, it's the transfinites. They're moving toward us.''

You turn and see Apollo shuffling about on the marble floor. His cloak of colorful flesh makes him look like undulating, psychedelic jellyroll.

"Oh no!" Mr. Plex screams. "Apollo is Pythagoras' favorite god. What if he sees this?''

A few crustaceans, the keepers of time, begin to repair a crack in the floor beneath Apollo's massive feet.

You grab Theano's hand. "Quick, back to the storage closet." You turn to Mr. Plex. "Stay here and scare Pythagoras away.''

"Sir, I can't do that.''

Pythagoras' dark watchful eyes miss nothing. "Theano!" he screams.

You gaze at Pythagoras' huge kithara. He seems to be running toward you. "Mr. Plex, do something!''

"Yes, sir.''

Mr. Plex rises up on his hindlimbs and stretches himself like Hercules unchained between two Temple columns. The sunlight reflects off his diamond body as his abdomen pulsates and his diamond teeth sparkle. He tosses back his head and gives a lion-like roar.

THE HISTORY BEHIND THE SCIENCE FICTION

Turks and Christians

Theano was interested in why there was conflict between the Christians and the Turks. Moslems and Christians have had a long history of discord in the Near East, including the

Crusades where European Christians came to the Holy Land in an attempt to take it away from the Moslems. Conflict lasted for centuries.

One famous example of strife between Turks and Christians occurred in the middle 1400s. In the year 1451, Mohammed II, surnamed the Conqueror, came to the Ottoman throne at the age of 21. In June 1452, he declared war on Constantinople and attacked with 140,000 men. His Moslem comrades were inspired by the belief that to die for Islam was to win paradise in the afterlife. Interestingly, Mohammed had hired Christian gunsmiths to create for him the largest cannon known to humanity: it hurled stone balls weighing 600 pounds.

Emperor Constantine, himself a Christian, led the defense of Constantinople with 7000 soldiers. His brave brethren had lances, small cannons, bows and arrows, torches, and crude firearms. Alas, his city walls were eventually destroyed. On May 29, the Turks fought across a moat filled with fellow Moslem soldiers and began to terrorize the city. Constantine himself was in the thick of the battle. When he finally was surrounded by the Turks, he screamed out, "Cannot there be found a Christian to cut off my head?" He continued to fight but disappeared and was never heard from again.

Once they had conquered Constantinople, the Turks massacred thousands of Christians. Nuns were raped. Christian masters and servants were taken as slaves. The St. Sophia church was transformed into a mosque and all its Christian insignia removed.

The fall of Constantinople and transfer of power from Christian to Islamic rule upset every throne in Europe. The wall between Europe and Asia had fallen. The papacy, which had always hoped to win over Greek Christianity, now saw the rapid conversion of millions of southeastern Europeans to Islam. Trade routes, traditionally controlled by the West, were now in "alien" hands.

All of this had a profound effect on mathematics. In fact, the fall of Constantinople to the Islamic Turks actually caused mathematics to flourish. The migration of Greek mathematicians to Italy and France was accelerated. Vasco de Gama and Columbus stretched the confines of civilization, and Copernicus stretched the heavens. Centuries of asceticism were forgotten in an explosion of art, math, and poetry, and pleasure. As Will Durant notes in his book *The Reformation*, the western migration of Greek scholars to Europe "fructified Italy with the salvage of ancient Greece." Religious war altered the landscape of mathematics and science forever. And the modern world was born.

數 數 數 數

Although the decimation puzzle in this chapter may appear to be simple from a mathematical standpoint, these kinds of problems have been studied for centuries. Perhaps more notable is that these problems have been tied to religious quarrels. For example, the "Turks and Christians" puzzle is presented in various literature, although I do not know from where (and when) it ultimately originates.[1]

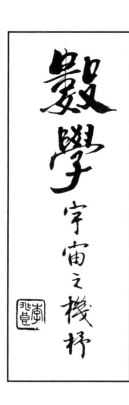

CHAPTER 9

The Ars Magna of Ramon Lull

I can't help thinking that scientists who write books about God and the Universe would be better advised to write more scientifically, and less poetically.
— Ralph Estling, *The Skeptical Inquirer*, 1993

I was astonished to discover how many of my close scientific colleagues practice a conventional religion.
— Paul Davies, *The Mind of God*

From beneath Hercules' thin, drooling mouth, a hundred throat appendages quiver aperiodically. The statue's several feet resemble horseshoe crabs, and its toes look like slugs.

Theano motions to the storage room door. "Don't go out there again," she says, her face clouding. "Not ever. It'd be asking for trouble."

You are sitting next to Theano on a straw mat on the floor of the temple's storage room. "You're right, it's too dangerous. Pythagoras may believe Mr. Plex to be an evil god who deserves destruction, or perhaps Pythagoras is jealous of the time you are spending away …"

Mr. Plex's right forelimb twitches uncontrollably. "Sir, thank goodness she's finally knocked some sense into you. I–I scared Pythagoras away for the moment, but I've a feeling he'll be coming back." Mr. Plex sounds out of breath.

You nod. "Next time I'll handle the situation myself. It wouldn't be good to have them start a religion based on Mr. Plex's peculiar physiognomy. In any case, there's nothing to worry about if we stay in this closet and watch the view-screens."

Theano gestures to the walls. "What's with the *five* view-screens?"

"I've positioned electronic flies at the north, south, east, and west sides of the temple. Pythagoras won't be able to return here without us seeing him approach."

Theano rises and stalks to the corner of the room. You have never seen her behave in quite this manner. In a moment she reappears. "What good is us *seeing* Pythagoras approach? He'll find us eventually."

You shake your head. "We can protect ourselves if necessary."

Mr. Plex looks at the view-screens. "Sir, why do you have *five* view-screens?"

"It's an experiment, Mr. Plex. I've sent a fly through time to the 13th century." You point to the fifth screen. "Today we'll talk about the *Ars Magna* of Ramon Lull."

"Excellent!" Mr. Plex screams.

Theano places her hands on her hips. "How can you have one of your teaching sessions at a time like this. Pythagoras could return any minute."

"Dear, just—"

"I'm not your 'dear.' "

Everyone is quiet. No one moves.

Mr. Plex ambles closer. "Ma'am, he meant no disrespect."

You nod.

Theano sighs and kicks at a cheeseburger wrapper.

During your last few days together, Theano has become more proficient in English with all its idioms. At first you were merely fascinated by her, charmed by her. Then you felt an emotional connection that appeared to be growing deeper with each day. Now you are desperate to teach her all of the profundities of life, to make her share your passion for mathematics, to make her love your books, art, music—to know how to fend for herself in the 21st century without being harmed. Still Theano keeps just beyond your emotional reach. In the next few days, you hope that some of the distance between you and her will melt away.

You shake your head to clear your thoughts. "As I was about to say, our time-fly is circling Ramon Lull, a Spanish theologian born in the year 1234."

Mr. Plex grins. "Four consecutive digits."

You nod as a noisy image of a man with a long white beard appears on your screen. He stands on the top of a large saddle-shaped mountain and gazes up towards the heavens.

"Sir, what's he looking at?"

You send a command to the electronic fly, and it points its head upward.

The sky is a perfect china blue, and great curling clouds race by like sentient sailing ships in a gust of wind. The fly returns its view to Ramon Lull.

"Sir, he doesn't look too good."

You nod. "The year is 1274. The place is Mount Randa on the island of Majorca. For several days he's been fasting and meditating. According

The nine steps leading to the Heavenly City by Ramon Lull. (From the *Liver de ascensu.*)

to historical records, he'll soon experience a vision where God reveals secrets to confound infidels and prove the certainty of religious dogma. Our history books also tell us that Lull will have a vision of the leaves of a lentiscus bush becoming mysteriously covered with letters from different languages. These different languages are the ones in which Lull's Great Art are destined to be taught. Soon after he has these visions, Lull will write the first book of many called the *Ars Magna*." You say the words *Ars Magna* in a hushed voice. "The book describes how to use Lull's mathematical methods for seeking the ultimate truth of the universe."

Theano wanders closer and appears interested. Good. She cannot resist the lure of your lesson. Perhaps any anger she has felt will dissipate like fog in the warm sun.

"Sir, did Lull's contemporaries see him as a genius or a lunatic?"

"Lull started his life with great hedonistic indulgence similar to St. Augustine, and he had a passion for married women. Nonetheless, Giordano Bruno, the great Renaissance martyr, considered Lull divine. Bruno thought Lull's work contained a universal algebra by which all knowledge and metaphysical truths could be understood."

You begin to sketch on the temple wall. "Lull believed that in every branch of knowledge there are a small number of simple basic principles, and by exhausting all possible combinations of these principles, we can explore all knowledge. To help Lull arrange and study all possible combinations of words, phrases, or ideas, he designed various geometrical figures and wheels. For example, we can arrange a set of words in a circle and connect one to another with lines."

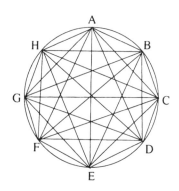

"Lull's most famous word organizer consisted of concentric circles which rotated about a central axis. The letter A representing God was often placed at the center of the circle. Around the circles were divine attributes such as goodness, greatness, eternity, and so on. By connecting the various attributes, we can obtain various permutations of concepts that stimulate the mind. For example, we can meditate on the composite idea that God is infinitely just. Lull even wrote a book describing how preachers could use his concentric wheels to arrive at stimulating topics for sermons."

Theano begins to scribble various diagrams on the wall. She grins. "Sir," she says imitating Mr. Plex's twangy metallic voice, "how could this help him learn anything about the universe?"

You smile. "Lull believed that each branch of knowledge rested on a few principles that formed the structure of knowledge in the same way that geometrical theorems are formed from basic axioms. By using his wheels to produce all combinations of principles, one explores all possible structures of truth and obtains universal knowledge."

Mr. Plex comes closer. "Seems fairly useless."

"Not entirely. The Lullian principle of randomly combining words to stimulate the mind is useful in producing startling fictional plots and verbal imagery that might not be considered otherwise. Meditating on bizarre combinations leads your mind to offbeat paths. In the 20th century, someone even once marketed a device to fiction writers called a "Plot Genii." Using this device, a hopeful author turns concentric circles to produce different combinations of plot elements."

You turn toward Mr. Plex. "Care to do a little experimenting on the computer?"

Mr. Plex nods. Theano's left eyebrow raises.

"You can create your own Lullian computer poetry generator by randomly selecting words and phrases which are then placed in a specific format, or 'semantic schema.' The program starts by reading 30 different words in five different lists (or categories), and stores these words in the program's memory arrays. The five categories are: adjectives, nouns, verbs, prepositional phrases, and adverbs. The words are chosen at random and placed in the following *semantic schema*, or minor variations of this schema:

Poem Title: *A (adjective) (noun 1)*
A (adjective) (noun 1) (verb) (prepositional phase) the (adjective) (noun 2).
(Adverb), the (noun 1) (verb).
The (noun 2) (verb) (prep) a (adjective) (noun 3).

The fact that "noun 1" and "noun 2" are used twice within the same poem produces a greater correlation—a cognitive harmony—giving the poem more meaning and solidity."

"Sir, let me make a list of words, and write a computer program."

Theano tosses Mr. Plex a personal computer, but her aim is significantly worse than yours, forcing Mr. Plex to make a leaping dive for the computer. He catches it but, in doing so, crashes into a marble frieze of representations of harpys, centaurs, Albert Einstein seizing Hercules, and of men feasting running about the entablature. Both Einstein and Hercules crash down upon Mr. Plex.

Mr. Plex struggles to free himself of them. "Did the transfinites warp Aphrodite into Einstein?" He shakes his head. "Never mind." Mr. Plex then furiously types on the computer's keyboard with his multiple legs.

The computer program asks for a number to seed a random number generator, and also for the number of poems to be generated:

```
> Please enter a seed for the random number generator > 81557
> How many poems would you like? > 5
```

When Mr. Plex presses the computer's Enter key, the computer prints five poems. He hands you a computer printout:

A CHOCOLATE PROPHET
A chocolate prophet drools while painting the glistening avocado.
Sensuously, the prophet gyrates.
The avocado chatters near a crystalline jello pudding.

A SEXY DREAMER
A sexy dreamer flies while eating the moldy spine.
While gradually melting, the dreamer squats.
The spine squats deep within a vibrating ocean.

A SKELETAL LIMB
A skeletal limb sings while dismembering the golden unicorn.
Vigorously, the limb smiles.
The unicorn screams above a lunar brain.

A CRYSTALLINE INTESTINE
A crystalline intestine regurgitates while listening to the moist dream.
Hideously, the intestine collapses.
The dream evaporates while grabbing at an apocalyptic sunset.

A BURNING MOUNTAIN
A burning mountain withers while dreaming about the glistening prophet.
With great deliberation the mountain makes love.
The prophet laughs while grabbing at a glittering snail.

"Sir, these poems don't impress me." He pauses. "But they sound more sophisticated than a simple random number generator could produce. All I did was give it thirty different words for each category."

There is an odd mingling of wariness and amusement in Theano's eyes. "Here let me try that," Theano says. She types in a different number for the random number generator.

```
> Please enter a seed for the random number generator > 90210
> How many poems would you like? > 10
```

She presses the computer's Enter key, the computer prints ten poems, and she hands you a computer printout:

A MAGNETIC BONE
A magnetic bone thrusts while puffing the frost-encrusted jello pudding.
While feeding, the bone wanders.
The jello pudding chatters while grabbing at a moldy knuckle.

A HALF-DEAD TOWER
A half-dead tower phosphoresces close to the delicious tongue.
While feeding, the tower smiles.
The tongue shakes at the end of a moldy robot.

A DYING TORCH
A dying torch chews inside the happy avocado.
Blindly the torch screams.
The avocado gyrates in synchrony with a moist centipede.

A CHOCOLATE VACUUM TUBE
A chocolate vacuum tube oscillates while massaging the frost-encrusted goose.
Grotesquely, the vacuum tube makes love.
The goose oozes simultaneously with a percolating sunset.

A CRYSTALLINE EARTHWORM
A crystalline earthworm jumps inches away from the sleep-inducing knuckle.
Hideously, the earthworm buzzes.
The knuckle regurgitates below a hungry web.

A FRIGID SUNSET
A frigid sunset cries before the dying intestine.
With all its strength, the sunset sings.
The intestine wanders while grabbing at a moldy goose.

A GOLDEN SAPPHIRE
A golden sapphire evaporates close to the moist mountain.
With great deliberation, the sapphire chatters.
The mountain gasps near a moldy brain.

A FAIRYLIKE FLAME
A fairylike flame sings inches away from the dying avocado.
While waving its tentacles, the flame undulates.
The avocado trembles while sterilizing a shivering tongue.

A DYING AVOCADO
A dying avocado regurgitates beneath the vibrating cloud.
Before dying, the avocado regurgitates.
The cloud drools a million miles away from a black sunset.

A PIOUS MAGICIAN
A pious magician implodes while rubbing the green veil.
Happily the veil shines.
The magician sings in synchrony with a skeletal diamond.

You study the poems. "Mr. Plex, your choice of words for your word lists is a bit odd."

"That's an understatement," Theano interjects.

"But you now see the Lullian concept in action. Even with a random choice of words, the computer creates simple poems. With a little more sophistication of rules, the poems might be quite thought-provoking." You reach for the computer. "Here, hand me that. I have my own word lists to try. The words come from Edward Fitzgerald's translation of the Rubaiyat of Omar Khayyam."

You type in the new word list and have the program print 10 new poems:

AN UNBORN SPANGLE OF EXISTENCE
An unborn spangle of existence ascends toward heaven and hell.
While transmuting, the spangle of existence jumps.
Heaven and hell weep while seeking their external destiny.

A BURIED IDOL
A buried idol evolves inside the divine nightingale.
In mind-inflaming ecstasy, the idol burns.
The nightingale dreams within annihilation's waste.

A CONSCIOUS CLAY
A conscious clay phosphoresces while pulling apart its shivering destiny.
Gazing longingly into space, the clay regurgitates.
Destiny evolves at the tip of heaven and hell.

AN UNBORN HYACINTH
An unborn hyacinth thrusts before the everlasting penalties.
With great satisfaction, the hyacinth sings.
The everlasting penalties evolve while listening to a shivering vessel.

A FROST-ENCRUSTED BIRD OF TIME
A frost-encrusted bird of time breathes at the end of a loaf of bread.
Hesitantly, the bird of time gyrates.
The bread cries while staring at a skeptical lamp.

A SKELETAL MORNING OF CREATION
A skeletal morning of creation oozes on the unborn seventh gate.
With great speed, the morning of creation laughs.
The seventh gate disintegrates while dreaming about a phantom annihilation.

A DYING PARADISE
A dying paradise vanishes at the end of the praying angel.
Gazing longingly into space, paradise burns.
The angel shivers while pulling apart a hungry loaf of bread.

TRANSLUCENT PROPHETS
A translucent prophet dreams within the shining blossom.
Ever-so-slowly, the prophet undulates.
The blossom wanders in spite of a shining key.

AN ETERNAL RUBY
An eternal ruby dissolves besides the invisible Sultan.
Erotically, the ruby makes love.
The Sultan regurgitates near a glowing throne.

AN INFINITE SWORD
An infinite sword gasps while rejecting the divine revelation.
While shivering, the sword evolves.
The revelation sighs while touching an unborn universe.

"Interesting," Mr. Plex says.

You nod. "Let me tell you about the mathematical side of Lull. He was a mathematician of sorts, although not a very good one." You go over to a wall and begin to draw. "Here's my favorite Lullian diagram." You point to a rectangle filled with words and geometrical shapes (see below). "Lull used this figure to show how the mind can conceive of geometrical truths not apparent to the senses, and to prove that there is only one universe rather than many universes." You pause. "The diagram at the middle of the bottom is of greatest interest. I'll redraw it more neatly." (see p. 111).

Theano and Mr. Plex lean closer.

"To produce the figure, we first inscribe and circumscribe a square about a circle. Next draw a third square, called the Lullian square, midway between the other two squares. By midway, I mean that the distance of the top of the outer square to the top of the midway square is equal to the distance from the top of the midway square to the top of the inner square. What can we know about the area and perimeter of this third square?" You pause. "Ramon Lull asserted that the third square has a perimeter equal to the circumference of the circle as well as an area equal to the circle's area."

Theano's fine, silky eyebrows rise a trifle.

Mr. Plex is gesturing wildly to the view-screen, but you pay no attention, wishing to finish your thought.

"Today, we know this is not true, that Ramon Lull was wrong about the areas and perimeters. However, is there *anything* we can say about the area and perimeter of this third square in relation to the circle?"

"Sir, look at the view-screen."

Diagrams used by Renaissance Lullists.

You turn and see a cloud of dust appearing on your right screen.

"I can't quite see what it is."

"Mon Dieu!" Mr. Plex screams.

On the view-screen are ten Greek warriors coming toward the temple. They are clad in magnificent garments of war—golden, studded harnesses and bronze helmets. At their sides dangle their traditional weapon of war, the long sword. They ride massive, black stallions. From their mounts hang daggers, sabers, rapiers, scimitars, wide-blade knives called misericords, and various other instruments of mayhem, the function of which you cannot quite discern.

They ride in perfect formation, each rider located at a point in four rows forming the tetraktys.

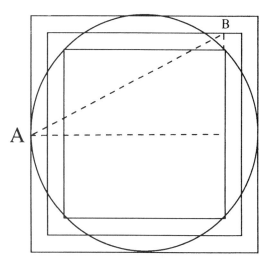

Enlargement of bottom, middle of figure on page 110.

"Let me handle this," you say.

Theano pulls on your arm. "You promised me you wouldn't go out there."

"Theano, I'm sorry. This will take a minute."

You open the backroom door and Theano follows. Mr. Plex remains behind, his forelimbs and abdomen shaking like an untuned engine.

"Maybe they're Pythagoreans," Theano whispers.

"Maybe Pythagoras hired them?"

"What do we do now?" Theano asked. Her whole body tightens, and then she takes a breath.

"Let's wait and see what happens." You feel a creeping uneasiness at the bottom of your heart.

When the riders are in earshot, you adapt your speech and body language to their inflated style and manner.

"From whence come these noble warriors?!" you cry. You try not to show any fear.

Theano stared up at the riders. Waiting. Tense. "What in hell are you talking about?" she whispers as she kicks you in the shin.

"Play along with me."

The large warriors remove their helmets and dismount. Five of the riders break from the tetraktys formation and gallop away into the distance. Could it be that they wish to report your location to Pythagoras? Or do they simply feel that five soldiers are sufficient to deal with you and Theano?

The tallest rider comes closer.

Standing before you is a huge muscular man with a dark beard. His eyes almost seem to glow a fierce red as he shouts something that is probably a variant of ancient Greek.

"Can you understand him?" Theano says.

"Sort of. I studied Greek dialects while on my museum ship in the future."

The warrior's grey-green eyelids droop and ooze a tiny amount of liquid which smells like absinthe. "ποινη? (What dost thou want in Our Kingdom?)" he says in a voice several octaves too low.

You attempt to mimic his ostentatious manner of speech. "I seek wise Pythagoras, geometer extraordinaire, possessor of all knowledge."

"Then you will seek him forever, Warrior. Leave whilst thou are able!" He threatens you and Theano with his long sword. The voice of the warrior is eerie, like a low-pitched wind whistling through a window on a lonely January night.

"And what of the pre-pubescent tart?" taunts another warrior as he lasciviously leers at Theano. You clench your fists so that your fingernails dig painfully into your palms.

Suddenly the two warriors remove a peculiar looking weapon from their cloaks and knock you into a lentiscus bush.

Since you generally avoid bringing advanced weapons back in time, the long sword is your weapon of choice in ancient Greece. This is not too much of a liability since gunpowder doesn't yet exist. In one way this restriction is a welcome one for you. You are an expert using both the English cutlass and the French cup hilt rapier. To Greece you have brought a beautiful replica of a 17th century English cutlass with a quillon to protect your hand. You fashioned the blade yourself using bundles of iron strips, hammered together, which you cut and bent repeatedly. Subsequent carbonization of the metal improved its strength and temper.

You disentangle yourself from the lentiscus bush and shout back in their archaic style of speech. "Thou darest challenge me?" It is time to bring out the heavy artillery. You reach into a thin chamber you have previously hollowed out in a column of the Temple.

Whenever you're threatened, you seem to go haywire and lose control. For some reason, you aren't afraid. Perhaps it is your martial arts training. Perhaps you are a little bit crazy.

One of the warriors kicks you in the knee, and the warrior blood within you rises to fever pitch. Your hand yearns for the touch of your sword, yearns for the glory of battle. You hold your sword above your head.

"τετρακτοζ!" you scream to assuage your humiliation. You grin and then lunge with the fury of a rabid wolf.

"Theano, stay behind me," you shout.

You lunge at all five of the warriors, your ever-moving blade producing a wall of flashing, cutting steel through which no warrior dares transgress. One warrior comes a little too close, and you stab him in the chest. Another throws his sword at you. You dodge it and then stab him in his arm. You grow weak, but you fight on. You've always loved a good fight.

The third warrior stares at you but does not move. It appears that he is assessing your fighting skills. He snarls something which roughly translates to "pretty damn good." Then he runs toward you. Your sword slashes, thrusts, and parries his every attack. He is about to bring his sword onto your head when you slip under him and smash your fist into his groin. A second later, you run your sword through his neck.

"Watch out!" Theano cries. She points to the leader who wields a double-bladed sword shaped like the Greek letter π. You have heard stories about these terrible instruments of mutilation and death. Ordinarily, when one sword hits another sword the vibration spreads out to affect the entire sword. With π swords, some vibrational modes are trapped within a branch of the π shape, and thus the double-blade damps vibrations, making the sword quiet and sturdy. When the warrior strikes your sword you hear no loud metallic ringing sound. Just silence. It is unnerving. The warrior notices your consternation and smiles. With lightning speed, you smash his body with your arm while at the same time throwing a piece of lentiscus bush at his head. He stops smiling.

"Arrrrrr!" screams the warrior. His long nose is pinched and white with rage. You scream back at him with a furious voice. Then you run toward him. At last, with one ruthless disengagement and thrust, you kill the leader of the five warriors.

You look at the one remaining warrior. "I don't know who is uglier, you or Medusa." You laugh at your own joke with a judicious amount of false gusto.

The warrior roars and attacks you. His strength is so great that the strike of his sword splits your sword into two useless pieces. You fall to the ground. If only you had your other sword hidden back in the storage room. You reach for the puny dagger in your belt. The warrior catches hold of your wrist, his rough fingers forming a muscular bracelet of incredible strength. You hear Theano gasp and see her run to the warrior. Then she gives a swift, vicious kick to the warrior's hip. The warrior loosens his grip.

You are about to take a deep breath of relief when the warrior's sword rips into you. The pain is so devastating that you drop the dagger, feeling pain like nothing you have ever known. You think of Theano, your growing affection for her, and you tell yourself you have to try to protect her, and that you aren't going to die, aren't going to die. …

The huge warrior comes at you again with his long sword. Without warning, you launch your foot up at his throat with a powerful karate kick. He is stunned and drops his sword. With a quick intake of breath like someone about to plunge into Arctic air, you quickly grab his sword, lunge forward, and remove his head.

THE SCIENCE BEHIND THE SCIENCE-FICTION

The Life of Ramon Lull

The character of Ramon Lull (1234–1315) is, of course, real. In fact, in his *Book of Contemplation*, Lull seeks to prove all the major tenets of Christianity. In this book, he is preoccupied by number symbolisms. For example, the work is divided into five books to symbolize the five wounds of Christ. Forty subdivisions symbolized the 40 days Christ spent in the wilderness. One of the 366 chapters is to be read each day, and the last chapter read only in leap years. Each chapter has ten paragraphs (the ten commandments), and each paragraph has three parts (the trinity), making a total of 30 parts per chapter (the 30 pieces of silver). Geometrical objects, such as circles, are often introduced as metaphors.

Lull often used letters to stand for words and phrases so as to condense his arguments into an algebraic form. For example, here is part of a paragraph from Chapter 335:

> Since diversity is shown in the demonstration that the D makes of the E and the F and the G with the I and the K, therefore the H has certain scientific knowledge of Thy holy and glorious Trinity.

Martin Gardner, who has written extensively on Ramon Lull's work, notes of the *Ars Magna*:

> There are unmistakable hints of paranoid self-esteem with the value Lull places on his own work in the book's final chapter. It will not only prove to infidels that Christianity is the one true faith, he assures the reader that he has "neither place nor time sufficient to recount all the ways wherein this book is good and great."

As you mentioned to Theano and Mr. Plex, Lull used many concentric wheels to examine universal truths. For example, one of Lull's wheels is concerned with: God, angels, heaven, man, the imagination, the sensitive, the negative, the elementary, and the instrumental. Another is used to ask questions such as: what? why? how great? when? where? how? etc. In his books *Ascent and Descent of the Intellect* and *Sentences*, he uses the rotating circles to formulate questions such as:

1   Where does the flame go when a candle is put out?
2   Why does the rue herb enhance vision while onions weaken vision?
3   Where does the cold go when a stone is warmed?
4   Could Adam and Eve have cohabited before they ate their first food?
5   If a child is slain in the womb of a martyred mother, will it be saved by a baptism of blood?
6   How do angels communicate with one another?
7   Can God make matter without form?
8   Can God damn Peter and save Judas?
9   Can a fallen angel repent?

In his book, *Tree of Science*, Lull asks over 4000 similar questions! Sometimes Lull gives answers to the questions, while other times he simply poses the question and lets the reader formulate a response.

Frequently Asked Questions (FAQ)

Below is a list of questions I frequently receive regarding Ramon Lull, followed by my brief answers.

1   *Did Lull have other outlets for his creative output?* Yes. Lull was a poet, and his collection of poems on *The Hundred Names of God* is one of his best known works.
2   *Did Lull believe one could communicate with the dead?* Yes. He rejected geomancy (divination using signs derived from the earth) but accepted necromancy (the art of communicating with the dead). Lull even used the success of necromancers as a proof of God's existence.
3   *Did Lull belong to a specific religious order?* Yes. While in a Dominican church he saw a brilliant star and heard a voice speaking from above: "Within this order thou shalt be saved." Unfortunately, the Dominicans had little interest in his art, and Lull therefore joined the Franciscans who found it of some value.
4   *How did Lull die?* At the age of 83, Lull set sail for the northern coast of Africa. By this time he had a long white beard. In the streets of Bugia, he started to shout about the errors of the Moslem faith and was stoned by an angry mob. Before he died, Lull is said to have had a vision of the discovery of the American continent by a merchant. Lull's remains now rest in the chapel of the Church of San Francisco at Palma, Spain, where they are revered as those of a saint.
5   *Does anyone follow Lull today?* Yes. Interest continues in Lull and his work. For

example, the Second International Lullist Congress took place at El Encinar and
Miramar, Majorca, in 1976, bringing together famous Lullist scholars from around
the world.

115

The Ars
Magna of
Ramon Lull

The Lullian Square

Before you were interrupted by the Greek warriors, you started to tell Theano and Mr.
Plex about a figure where a square is inscribed and circumscribed about a circle (p. 111). As
mentioned, Lull claimed that a third square positioned midway between the two original
squares has a perimeter equal to the circumference of the circle as well as an area equal
to the circle's area.

Could Lull have been correct?

If we let R be the radius of the circle, then the outer square has sides of length $2R$ and
an area of $8R^2$. The inner square has sides of length $\sqrt{2}R$ and an area of $2 \times R^2$. The square
in the middle has a side which is the average of the inner and outer square and therefore has
a side length of $(2 + \sqrt{2})R/2$, and an area of $(3 + 2\sqrt{2})R^2/2$ and perimeter of $(4 + 2\sqrt{2})R$.

We can compare the area and perimeter of the Lullian square with the circle by
assuming $R = 1$:

	Circle	Lullian Square
Area	3.14	2.91
Perimeter	6.28	6.82

A circle encompasses more area than any other shape of equal perimeter length. In
other words, you would need less fence to enclose a circular piece of land than a square
piece of land.

Jeff Vermette from Rockville, Maryland, points out that any geometric figure that has
both the same perimeter and area of a given circle must be a circle itself. Therefore, the
Lullian square could not have had the same area and circumference as the circle. The ratio
between the area and the perimeter is maximized when the figure is a circle, so any figure
whose ratio equals the ratio for a circle must itself be a circle. This elegantly shows the
invalidity of Lullian mathematics without having to resort to calculators and three significant digits to demonstrate the fallacy.

Martin Gardner, in *Science: Good, Bad and Bogus*, notes that if a diagonal line AB is
drawn on Lull's figure, as shown on page 111, it gives an amazingly close approximation to
the side of a square with an area equal to the area of the circle. Why is this so?

Computer Poetry

In this chapter, we've also discussed the Lullian approach to generating computer
poetry, a subject I cover in my book *Computers and the Imagination*. Experiments with the
computer generation of poetry, Japanese haiku, and short stories provide a creative programming exercise for both beginning and advanced students. Computer-created poetry
and text also provides a fascinating avenue for researchers interested in artificial

intelligence—researchers who wish to "teach" the computer about beauty and meaning. You may want to read about early work in the computer generation of poetry in Riechardt's book, *Cybernetic Serendipity*. Other past work includes the book *The Policeman's Beard is Half Constructed*—the first book ever written entirely by a computer. The program which generated the book was called RACTER and was written by W. Chamberlain and T. Etter.[1] Finally, there is the Kurzweil Cybernetic Poet, a computer poetry program which uses human-written poems as input to enable it to create new poems with word-sequence models based on the poems it has just read. When these computer poems are placed side by side with human poems, many humans can not judge which poems were made by humans and which by the Cybernetic Poet.

You will also find that computer-produced texts are a marvelous stimulus for the imagination when you are writing your own (non-computer generated) fictional stories and are searching for new ideas, images, and moods. If you are a visual artist in search of new subject matter, computer-generated poems can provide a vast reservoir of stimulating images.

If your computer has access to a thesaurus, you may induce further artificial meaning in your computer-generated poems. This is accomplished by using the thesaurus to force additional correlations and constraints on the chosen words. For example noun 1, noun 2, and noun 3 in the semantic schema would all appear within the same thesaurus entry. You should also design your own semantic schema and use additional word lists. Try using probability matrices to produce the poetry. The matrices would consist of an array which makes your program more likely to pick certain combinations of words. Some entries in the matrix may be zero, which would disallow impossible combinations of words.

Lullian Approach to Music Generation

You may occasionally encounter natural scenes that remind you of a painting, or episodes in life that make you think of a novel or a play. You will never come on anything in nature that sounds like a symphony.
—Martin Gardner, *On the Wild Side*, 1992

In my book *Mazes for the Mind: Computers and the Unexpected*, I discuss Lullian methods for generating melodies. In particular, I describe the fictional work of Professor Mutcer of the Electrical Engineering Department at Harvard University who decided to build a musical melody generator that would continuously produce different 50-note melodic progressions. The melody machine would generate one melody after another, selecting for each melody a different combination of notes from the piano keyboard. His machine consisted of 88 oscillators, each of which produced a single tone. A random number generator was used to select which of the 88 oscillators were playing at any particular moment. The machine played the 50 random notes, one at a time.

Other versions of his music machine generated all possible 50-note melodies by sequentially trying all 88 tones for the first oscillator, while keeping all the others constant, and then stepping each oscillator sequentially (something like an odometer on your car's dashboard). Here are the first few melodies this version of the machine produced for Professor Mutcer, starting with A, B, C, D, …:

```
A, B, C, D, E, F, G, … (First Melody)
B, B, C, D, E, F, G, … (Second Melody)
C, B, C, D, E, F, G, … (Third Melody)
D, B, C, D, E, F, G, … (Fourth Melody
    …                      (Etc.)
```

Notice that first the machine stepped through all possible notes for the first position in the melody, starting with the lowest note on the piano (A = 27.5 hertz). The first song that the machine produced using this approach (top line in the example) was simply a melody consisting of the first 50 white notes on a piano keyboard played in order from low to high pitch. In the second melody, the first A note has switched to B, and so on. Later in Mutcer's research, black notes were also included. Either of these versions of the music machine (i.e., random or "odometer" versions), could be built without much difficulty. The duration of each note (quarter note, half note, eighth note, etc.) could also be selected at random.

Professor Mutcer set the machine in action and began to listen to the endless sequence of different melodies that came from the music machine. Most of the melodies made no sense at all to his Western ear. They looked like this:

But since the music machine played *all* possible combinations of musical notes, Mutcer began to find some nice tunes among the senseless, junk melodies:

Mutcer reasoned that a careful search would also reveal every melody written by Michael Jackson, Madonna, Sheryl Crow, Yanni, Beethoven, Bach, the Beatles, and Bananarama.

The machine would also produce every melody that Madonna discarded in frustration in her plush and high-tech recording studios.

Mutcer's music machine would even generate every melody ever played since ancient humans blew on wooden flutes or on the horns of goats. Moreover the machine would play every popular tune in the future, every musical hit from the year 2200. Musical publishers having Mutcer's machine would simply have to sit and listen, and select the good songs from the gibberish—which they do daily anyway.

A day after he built the music machine, Professor Mutcer proudly showed the device to several of his graduate students. A week later, he instructed his students to plug themselves into the music machine every day and press a button to record a musical score whenever they heard a particularly interesting melody. A few machines were built, and students took turns listening. No machine was ever idle for more than a few seconds as one student replaced another at this listening task. Interestingly, after about an hour of listening, one student was rumored to hear several phrases from the beautiful *Moonlight Sonata*. After two weeks, Mutcer himself heard both the *Cantata No. 96 Aria Ach, ziehe die Seele mit Seilen der Liebe* by Bach and *Havah Nagilah* (Israeli Hora). Another student fainted when she heard a particularly powerful and hypnotic tune. Although she did not know it, the tune just happened to be a current best seller on a small red planet circling the star Alpha Centuri. Mutcer copyrighted the best of the new musical scores, which were soon bought by large music publishers in New York and Rio de Janeiro. Mutcer became a millionaire.

Lullian Generation of Book Titles

Book authors have used the Lullian approach to generate effective and marketable book titles. Nancy Kress, a Contributing Editor for *Writer's Digest*, suggests that authors create marketable titles for their own books by first jotting down all the significant words in their story or novel. These can be words that suggest the action (love, murder, death), the characters (Sophie, Lazarus, salesman, husband), their relationships (lovers, strangers, enemies), the setting in time or space (IBM, the Four Season's Restaurant, Summer, 1984), the theme (love, revenge), a motif (roses, vampires, cars, darkness), etc. Next, the words are arranged according to certain semantic schema. Here are some examples. (I would be interested in hearing from those of you who have found other best-selling titles using this approach.)

1 **Possessive-Noun**. Examples: *Rosemary's Baby* (Ira Levin), *Finnegan's Wake* (James Joyce), *Sophie's Choice* (William Styron), *Childhood's End* (Arthur C. Clarke), *Schindler's List* (Thomas Keneally).

2 **(Article)-Adjective-Noun**. Examples: *Jurassic Park* (Michael Crichton), *Spider Legs* (Piers Anthony and Clifford A. Pickover), *The Witching Hour* (Anne Rice).

3 **(Article)-Adjective-Adjective Noun**. Example: *Another Marvelous Thing* (Laurie Colwin).

4 **Noun-and-Noun**. Examples: *Pride and Prejudice* (Jane Austen), *Crime and Punishment* (Doestyevsky), *War and Peace* (Tolstoy), *Beggars and Choosers* (Nancy Kress).

5 **(Article)-Noun-of-Noun**. Example: *Death of a Salesman* (Arthur Miller)

6 **(Article)-Noun-for-Noun**. Examples: "Flowers for Algernon" (Daniel Keys), *Requiem for a Heavyweight* (Rod Serling).

7 **Prepositional Phrase**. Examples: *Out of Africa* (Isak Dinesen), *In Our Time*, *For Whom the Bell Tolls* (Ernest Hemingway), *Of Mice and Men* (John Steinbeck).

8 **Article-Noun-Prepositional Phrase**. Examples: *Catcher in the Rye* (J. D. Salinger), *The Bridges of Madison County* (Robert James Waller).

9 **Infinitive Phrase**. Examples: *To Kill a Mocking Bird* (Harper Lee), *To Dance With the White Dog* (Terry Kay).

10 **Adverbial Phrase**. Examples: *Where the Wild Things Are* (Maurice Sendak), *How to Win Friends and Influence People* (Dale Carnegie).

11 **The Noun Who**. Examples: *The Spy Who Loved Me*, *The Man Who Melted* (Jack Dann).

12 **A Command**. Examples: *Remember Me* (Mary Higgens Clark).

13 **A Sentence with a Key Story Idea**. Examples: "Can You Feel Anything When I Do This?" (Robert Sheckley), "I Have No Mouth But I Must Scream" (Harlan Ellison), "Repent Harlequin, Said the Ticktockman" (Harlan Ellison).

Although Nancy Kress suggests using this Lullian approach without the aid of the computer, I recommend that you write computer programs which dump lists of hundreds of titles in the same way that I have written programs to generate simple computer poetry. Scan through the list of computer-generated titles, pick the best, and go sell your novel!

Lullian Approach to Inventions

Where there is an open mind, there will always be a frontier.
—Charles F. Kettering

Those of you who are interested in computers or are creative engineers may like to use the Lullian approach for inventing new products and patenting the results. One way to stimulate your imagination is to have a computer program generate an invention title by randomly choosing from a list of devices and then also choosing from a list of features (Pickover, 1994). If you can think of suitable applications for the invention, it is relatively easy to embellish the basic concept suggested by the random title and generate patentable ideas using this approach. The list in Table 9.1 will help you understand this approach. Pick a device from the first column, add the word "with," and then choose a feature from the second column. For example, "Mouse with Infrared Security Alarm" might be the title of your invention. Think about all the ways this could be achieved and all the applications of the device. Have your friends add devices and features to your own list for more interesting patent ideas.

The Pi Sword

The π sword, carried by the Greek warrior, has loose underpinnings in physics which suggest that strangely shaped objects can damp sounds when struck, and can be particularly strong. For example, consider hypothetical swords with fractal edges such as those

Table 9.1. The Lullian Approach for Creating Patentable Inventions. Pick a Device from the First Column, add the Word "with," and Then Choose a Feature from the Second Column.

Device	Feature
Pen	LED Flasher
Clock	Bell
Key	Speech Synthesis
Mouse	Light Meter
Keyboard	Touch-Activated Switch
Joystick	Timer Plus Relay
Graphics Puck	Missing Pulse Detector
Trackball	Voltage-Controlled Oscillator
Terminal Screen	Frequency Meter
Pencil	Light Dimmer
Terminal Keys	Infrared Security Alarm
On/Off Switch	Analog Lightwave Transmitter
Dial	Protection Circuit
Remote Control	Adjustable Siren
Compass	LED Regulator
Level	Wrist Band Attachment
Screwdriver	LED Transmitter/Receiver
Watch	Speech Recognition
Cursor	Volume Control
Menu Icon	1-Minute Timer
	Dual LED Flasher
	Neon Lamp Flasher
	Solar Cells
	Dark-Activated LED Flasher
	Break-Beam Detection System
	Phone Activated
	Phone-Controlled
	Piezioelectric Buzzer
	Bargraph Voltmeter

seen in jagged Koch snowflake curves (Pickover, 1995). These kinds of swords may have unusual physical properties. In 1991, Bernard Sapoval and his colleagues at the Ecole Polytechnique in Paris found that fractally shaped drum heads are very quiet when struck. Instead of being round like an ordinary drum head, these heads resemble a jagged snowflake. Sapoval cut his fractal shape out of a piece of metal and stretched a thin membrane over it to make a drum. When a drummer bangs on an ordinary drum, the vibration spreads out to affect the entire drum head. With fractal drums, some vibrational modes are trapped within a branch of the fractal pattern. Faye Flam in the December 13, 1991, issue of *Science* (vol. 254, p. 1593) notes: "If fractals are better than other shapes at damping vibrations, as Sapoval's results suggest, they might also be more robust. And that special sturdiness could explain why in nature, the rule is survival of the fractal." Fractal shapes often occur in violent situations where powerful, turbulent forces need to be damped: the surf-pounded coastline, the blood vessels of the heart (a very violent pump), and the wind- and rain-buffeted mountain.

CHAPTER 10

Death Stars, a Prelude to August 21, 2126

Humility collects the soul into a single point by the power of silence. A truly humble man has no desire to be known or admired by others, but wishes to plunge from himself into himself, to become nothing, as if he had never been born. When he is completely hidden to himself in himself, he is completely with God.

—Isaac of Nineveh, 6th century

When beggars die, there are no comets seen: The heavens themselves blaze forth the death of princes.

—Shakespeare

You slam shut the door to the backroom hideout of the Temple of Apollo. In your hand is a bloody portion of your broken 17th century English cutlass.

"Let's get the hell out of here," you say.

Theano and Mr. Plex nod their heads as they gaze at a foot-high statue of Dionysus in a condition of mystic frenzy. Everything about the statue appears normal, except for the fact that Dionysus, the God of Wine, has a mass attached to his leg. It juts grotesquely from his body, like a deformed, rotting pear that has somehow taken root in his knee. You tap on the protruding marble globule.

Outside it is dusk, and the view-screens show dozens of chariots converging on the Temple of Apollo. They come from the North, South, East, and West. Riding in each chariot are two soldiers carrying shields and pikes. Apparently the fact that you killed several Greek warriors has led to an escalation of conflict. Not a good sign.

Theano comes closer to you executing a playful pirouette. "You were amazing with your sword. Quite brave!"

Mr. Plex remains standing with the ramrod posture of a British brigadier. "Sir, if I may say so, you took unnecessary risks."

You wave your hand. "It's over. It doesn't matter. My cuts don't seem to be bleeding much."

Theano tries to bandage your wounds. "Where does it hurt?" she asks.

"Ouch. Everywhere, but don't fuss."

"At least let me put some medicine on those cuts." She begins to apply special healing salves that should cause your torn muscles to repair themselves quickly while at the same time relieving you of pain. "Try to take it easy," she says.

You are pacing. "Tell me more about your husband."

"Like what?"

"Like why you are uncertain about staying with him?"

"Is that relevant?"

"Maybe."

"He wasn't cruel. And there wasn't another woman."

"Then why are you growing apart?"

"His arrogance bothers me. He thinks himself a god."

"But you loved him once?"

Theano sighs. "Yes, but he changed."

"How?"

"I was always very shy. Never thought myself pretty or witty enough."

"That's hard to believe."

"I had a horrible feeling of inferiority."

"Why?"

"My father. It's not easy..."

"What about Pythagoras?"

"He spreads a myth that his golden thigh means he is the son of Apollo or the god himself. He's too interested in numbers. Never a moment's relaxation. Maddening. I realized he was not just enthusiastic, not just an overachiever, but obsessive. Kept whispering that damn theorem of his in his sleep." She is silent for a few seconds. "I love mathematics. It's my passion. But he—he stopped caring about me."

You are quiet for a few minutes and then go over to the wall and press a button on the view-screen. It sends a signal to an electronic fly, causing it to look up at the darkening sky. "Just in time to see my favorite comet—Donati's comet."

On the view-screen is a comet. Its main tail is curved with two shorter, straight ones protruding from the sides.

"Beautiful," Theano says.

Mr. Plex takes a deep breath. "Will it hit us, sir?"

"No, it's too far away."

The three of you stare at the comet, and then, finally, you press a button on your waist, sending the view-screens away from ancient Greece. The backroom of the Temple of Apollo is now devoid of any signs of technologically advanced equipment. The computers are gone. The electronic flies are ordered to plunge into the Aegean sea—unwholesome treats for the sharks and the barracudas. The only futuristic object that remains on the floor of the temple's storage closet is a McDonald's hamburger wrapper from the year 2080 and a single, week-old French fry.

The floor begins to shake, and a small crack appears. A dozen crustaceans emerge from the crack and hastily begin to repair it.

"Not much time!" Mr. Plex screams.

You grab the mutant statue of Dionysus. "I'll take this with me. A memento of our journey to ancient Greece." You turn to Theano. "You sure you want to go with us?"

"I wouldn't miss it for the world." Theano smiles. It seems like an invitation to shared happiness.

"Sir, where are you taking us?"

"A surprise, Mr. Plex."

A statue of Hermes begins to buck like a bronco, and then slowly creeps toward the three of you. A tiny crustacean slithers from his belly button. From outside, you can hear the clattering of horses' hooves.

There are too many gods coming to life: Hermes, Aphrodite, Apollo, Diana. The transfinites really want you out of here.

Mr. Plex looks from one Olympic god to the next and quickly turns to you. "Sir, do you believe that there is a proof of God's existence?"

You are quiet. Theano stares at you, and you see something flickering far back in her eyes which streams like gold in the advancing darkness. You take a deep breath. Mr. Plex's massive abdomen begins to stretch like an accordion.

You raise your left eyebrow.

Theano gasps. "Can you tell us?"

You nod. "There are roughly ten thousand objects in outer space which are half a kilometer (a third of a mile) or more in diameter moving on Earth-intersecting orbits."

Theano runs her hands through her hair. "What's this have to do with God?"

Mr Plex's forelimb quivers. "Sir, that many outer space bodies? Where do they come from?"

"From the cold outer reaches of the solar system. Some are ancient comets; others come from the asteroid belt between Jupiter and Mars."

"If one were to strike the Earth?" Theano starts.

"It would cause more damage than all the world's nuclear weapons detonated simultaneously." You see confusion in Theano's eyes. "It would be devastating, more powerful than any earthquake, and it's only a matter of time before one slams into us."

"Sir, the Bible is loaded with predictions of destruction. From Revelations 16:18 … 'Then there came flashes of lightning, rumblings, peals of thunder, and a severe earthquake. No earthquake like it ever occurred since man has been on Earth … '"

You nod. "The huge mass of Jupiter can easily disturb individual asteroids' orbits, sending them plunging toward the Sun and threatening Earth. Comets originating in the mysterious Oort cloud, situated a light-year from our Sun, also pose a threat to us. So do rogue planets, black holes, and neutron stars."

Theano looks worried. "With so many threats, how can humanity survive?"

You move so close to Theano you can hear her breathe. "Some astronomers even believe that our Sun is part of a double star system, the second star of which is hidden."

"Hidden?" Theano asks.

"Double star systems are common in the galaxy. Some astronomers suggest that our sun has a companion star called Nemesis or the Death Star which is too dim and too far away to have been discovered yet. As it slowly orbits around the Sun, it periodically disturbs distant comets and may send some plunging to Earth. Geologists have found that devastating ecological destructions occur in a cyclical fashion—every 30 million years."

You pause and lower your voice. "The fact that humanity has not already been destroyed by the myriad projectiles around us suggests that we are protected, that God exists. At least some people think so. I call it the 'God protection-policy argument (GPPA).' "

Theano puts her hands on her hips. "So how do you think the world will end?"

"Despite the GPPA, I believe that Earth will eventually meet its doom when the comet Swift-Tuttle collides with the Earth on August 21, 2126. And we will be there minutes before it happens so we can learn how the world ends."

"Sir, is that where you're taking us now?"

There is a pounding on the door. Mr. Plex runs over and braces the door with his diamond body. Can he feel the pikes as they begin to tear through the door?

You come closer to Mr. Plex. "No, for now, our destination will be more serene."

You stoop down to examine a pike which protrudes 3 inches into the doorway. "Seems like I did not use sufficient fiberglass compound when I had the door made."

Mr. Plex puts his full weight on the door as the pikes continue to pierce it. "Uh, sir—"

You look into the eyes of Mr. Plex and Theano. "Say good-bye to ancient Greece."

"And to Pythagoras," Theano says.

You press a travel button on your belt. "Here we go again-n-n ..." You feel a sickening jarring motion. The odor of rose blossoms fills your nostrils. You, Theano, and Mr. Plex are whisked away....

THE SCIENCE BEHIND THE SCIENCE FICTION

Space—it seems there's never enough of it, even in space. Circling the sun along with Earth and the other planets are a swarm of asteroids and comets, many of them big enough to deliver a catastrophic jolt—which is what happened 65 million years ago, wiping out the dinosaurs, say many scientists.
—*Pittsburgh Supercomputing Center Newsletter*, 1994

There are stars in our universe which one cannot see.
—Jean Audouze

Swift-Tuttle

All the astronomical information that you gave Theano and Mr. Plex is based on our current scientific observations. There are indeed 10,000 outer space objects, nearly a kilometer in diameter, moving in Earth-intersecting orbits. The Death Star hypothesis is actually discussed by modern astronomers, and the comet Swift-Tuttle is real.

When the Swift-Tuttle comet came close to Earth in 1993, astrophysicists performed various computations which suggested the possibility that Swift-Tuttle will return and hit the Earth on August 21, 2126. If the comet collides with Earth, there will be a global catastrophe, and human civilization will end. Astronomers have since revised their initial prediction using improved orbital calculations, and now suggest that the comet will in fact miss our planet by two weeks. Could these latest computations be in error? Astronomical predictions are often difficult to make given the complexity of the calculations and the rate at which computational uncertainties grow over long periods of time. The orbit of comets is also difficult to predict accurately because a comet's orbit is often disturbed by numerous gravitational encounters as well as uneven jetting of material from its tail.

Doomsday by cometary collision is not a new area of speculation. For example, the Reverend William Whiston, a contemporary of Newton, predicted that the Earth would eventually be destroyed by collision with a comet. In 1970, a bright comet named Bennett caused alarm in Arab countries because it was mistaken for an Israeli weapon of war.

Some believe that the object which impacted in the Tunguska region of Siberia in 1908 was the head of a tiny comet (see Chapter 21). At 7:30 a.m., a huge flash lit up the sky, and an explosion, equivalent to 800 Hiroshima bombs, flattened the Siberian forest for hundreds of square miles. Luckily, the crash site was in an unpopulated forest. Recent analysis suggests that the culprit was a stony asteroid, about the diameter of a football field, that exploded five miles above ground.

When you were with Mr. Plex and Theano, you saw the comet Donati on your view-screen. Comet Donati, with its three tails, is real. In October 1858, the length of the main tail reached 80,000,000 km. The comet has an orbital period of about 2000 years.

The Cosmic Foul Ball

Could a large comet or asteroid plummet toward Earth—a cosmic foul ball from heaven? Many scientists believe this already happened 65 million years ago, spelling doomsday for the dinosaurs. Most scientists believe that it will happen again in the next 100 million years.

Even though we can't perform experiments to determine exactly what would happen to Earth when a collision of this kind takes place, something similar happened in the summer of 1994 to give us some insight. The planet Jupiter took the hit for us, giving scientists an unprecedented chance for observing doomsday from a safe distance. In fact, for months the comet Shoemaker-Levy 9 had doomsday written all over it as scientists predicted its trajectory toward Jupiter. Rather than a single body with a long tail, the comet was "squashed," composed of at least 21 fragments held together like a string of pearls.

Starting July 16, the train of comet fragments fell into the vast Jovian atmosphere, creating a fireworks show eagerly photographed by telescopes around the Earth and in space. The fireballs caused by many of the impacts were generally more energetic than expected. Some of the rising plumes could be glimpsed above the rim of the planet even before the point of penetration had rotated into direct view. According to the American Institute of Physics' experts, one of the fragments' explosions amounted to the equivalent of a 6 million megaton bomb. The plumes arising from the successive strikes persisted for days, gradually cooling and flattening into spots, some as large as the Great Red Spot, the famous large spot on Jupiter's surface.

In May 1996, just four days after it was spotted, an asteroid whizzed by Earth missing the planet by 280,000 miles—scarcely farther away than the moon, a hairbreadth in astronomical terms. About a third of a mile across, it was the largest object ever observed to pass that close to Earth. If it had hit the Earth, it would have struck at 58,000 m.p.h. The resulting explosion would have been like taking all of the U.S. and Soviet nuclear weapons, placing them in single pile, and blowing them up. If it had hit New York City or Washington, D.C., there wouldn't be anything left of those cities.

Dr. Oort

Comets are "official members" of our solar system. According to a theory developed by Jan Oort in 1950, there exists a diffuse reservoir of gas, dust, and comets which is gravitationally part of the solar system but located about 40,000 AUs from our Sun. (An AU, or astronomical unit, is the mean distance of the Earth from the Sun and equal to 149,597,870 km.) Nearby stars occasionally brush this cometary cloud, causing comets to fall toward the center of our solar system. (Dr. Oort estimated that there are 100 billion comets in what is now called the Oort cloud.) If a comet travels close to Jupiter on its way into our solar system, it may be thrown into a short-period orbit. Otherwise, it moves back toward its home in the Oort reservoir not to return to the neighborhood of the Sun for a very long time.

Interestingly, comets were once regarded as atmospheric phenomena. Most European observers agreed with Aristotle that comets were simply "exhalations" of the atmosphere.

2,100 Ra-Shalom

Most asteroids, or minor planets, lie between the orbits of Jupiter and Mars. By 1990 the number of named asteroids was over 3000. Scientists once believed that asteroids were confined to the zone between Jupiter and Mars, but in 1898, Carl Witt discovered "433 Eros," an asteroid which moves in an orbit approaching within 23 million km of Earth. In 1932, the asteroid 1,862 Apollo was discovered, and it actually crosses the orbit of the Earth. Earth-crossing asteroids are now collectively known as Apollos, and new Apollonian asteroids are discovered each year.

Asteroids whose orbits lie *within* the Earth's orbit around the Sun are part of the "Aten group" (pronouced "a-tuhn"), the best-known members of which are 2,062 Aten and 2,100 Ra-Shalom. Not many Aten asteroids are known, but they may be numerous. Authors Patrick Moore and Gary Hunt note, "There is always a chance that the Earth will collide with an Apollo or Aten asteroid, and indeed this must have happened in the past."

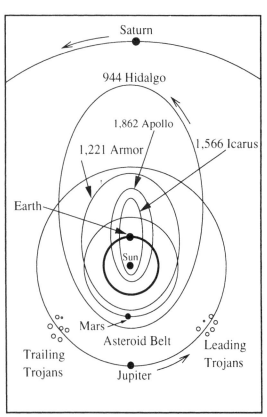

The orbits of several asteroids which leave the traditional asteroid belt between Mars and Jupiter and come close to Earth.

What Do Astrophysicists on the Internet Think of the GPPA?

In this chapter, we've encountered the "God protection-policy argument" (GPPA) which states, "Because humanity has not already been destroyed by the myriad projectiles around us, this means we are protected, and God exists."

After speaking with a number of astrophysicists on the Internet, we have come to the conclusion that the GPPA is not necessary to explain humanity's continued existence— because space is so vast, and objects so widely dispersed, a massive collision with Earth would normally be expected to occur roughly every 100 million years. The history of humanity is only two to three million years old. However, it is possible that our reptile ancestors suffered an impact 65 million years ago when they were probably destroyed by an asteroid of the 10 km size class. In 1980, Luis Alvarez, a Nobel laureate physicist, and his geologist son Walter concluded that a huge body struck the Earth at the end of the Cretaceous period, blanketing Earth with smoke and dust that blocked the sun for months and destroyed the dinosaurs.

The Alvarez theory seems to be gaining support from other scientists. For example, geologists have recently found a submerged crater off the Yucatan coast of Mexico and identified it as a likely site of the impact. From observing the crater, scientists judge the asteroid to have been 10 kilometers in diameter with an energy release equivalent to 200 million hydrogen bombs.

Interestingly, objects 10 meters in diameter collide with the Earth's atmosphere every year, but most explode high in the atmosphere. American military satellites have detected dozens of atmospheric explosions the size of the Hiroshima bomb during the last 30 years.

Virtually every astrophysicist with whom I spoke believes that humanity will eventually be visited by a large body, such as an asteroid or comet, and we will all be destroyed unless we can destroy the body before it hits Earth. It is only a matter of time....

What do Astrophysicists on the Internet think of Nemesis?

Some scientists believe the Death Star to be improbable and unnecessary to explain the historical record of impacts from objects in outer space. The Death Star was hypothesized at the time when scientists discovered frequently recurring extinctions in the history of life. Some researchers explained this by hypothesizing the existence of a binary companion to our Sun. The Death Star was thought to travel on a very elliptical orbit which frequently passed through the Oort cloud of comets, thus disrupting cometary orbits and sending many new comets into our inner solar system. However, these periodic animal extinctions were in actuality not very pronounced and regular; therefore, the Death Star is not needed to account for the available data on animal extinctions. Another reason that the Death Star hypothesis is not subscribed to by all astronomers is that the sky has been mapped very carefully, including very dim stars, and none of these stars is close enough to serve as a binary companion of the Sun. Some astronomers feel that if Nemesis were located sufficiently far away, it would eventually be pulled away by other stars in the Milky Way and thus would be short-lived (by geological standards).

The chances of finding Nemesis grow smaller by the years, since astronomical searches of the sky in the deep infrared frequencies haven't revealed Nemesis. Thus,

Nemesis, if it exists, must be very dark, not only in visible light but also in the infrared. Finally, since our Sun travels close to neighboring stars every few million years, there is no pressing need to postulate the existence of a binary companion which perturbs comets in the Oort cloud.

Still, astronomers admit the possibility of a Death Star. And they continually watch....

A Conversation with Bernard Peter Gore

I spoke with Bernard Peter Gore, author of *The Death of Nemesis* (1985), to obtain more information on the Death Star hypothesis. The idea began in early 1984 when authors David Raup and John Sepkoski suggested a probable cycle in mass extinctions of 26 to 30 million years. This initial report was rapidly followed by a theory proposing that the existence of a brown or red dwarf star as a distant companion to our Sun, in a highly eccentric orbit. At its perihelion (the orbital point where it is nearest the Sun), this companion disrupts the inner Oort cloud sending a shower of comets into the inner solar system, several of which impact the earth, causing damage to the environment and the extinction of species (Whitmir and Jackson, 1984).

In particular, Raup and Sepkoski studied marine fossil records and found an extinction period of 26 ± 1 million years (with a variance of 9 million years and a range of intervals between extinctions of 17–53 million years). They arbitrarily decided that a mass extinction occurred whenever 2% of the total number of species disappeared from the fossil record across the boundary of two geological strata. Researchers Michael Rampino and Richard Stothers used a figure of 10%, obtained a similar result, and proposed that mass extinctions are caused by comet impacts, based on two examples in which a layer of iridium exists in the rock strata close to an extinction event (Rampino and Stohers, 1984). The iridium was assumed to come from a comet or asteroid impact (Alvarez, 1980).

Bernard Peter Gore raises various doubts about the Nemesis hypothesis. For one thing, he feels that the extinctions started *before* the alleged impacts because significant faunal extinctions seem to have been well under way before the deposition of the iridium layer (Schwartz and James, 1984)!

Finally, in 1985, Antoni Hoffman showed that the periodicity claimed by Raup and Sepkoski relies on arbitrary dating decisions and, without these assumptions, the data is insufficient to support extinction periodicities. Raup and Sepkoski admitted from the start that the evidence might be insufficient to support the conclusions.

Incidentally, the Death Star was named after "Nemesis," the Greek goddess who relentlessly persecutes the excessively rich, proud, and powerful. Researchers Marc Davis, Piet Hut, and Richard Muller wrote in their 1984 paper, "We worry that if the companion is not found, this paper will be our Nemesis."

Computer Models of the GPPA

You can investigate the God protection-policy argument using a computer program. Place asteroids in elliptical Earth-intersecting orbits with random eccentricities. Set the Earth in orbit (see Program Code in the Appendix). How often does a collision take place?

For simplicity, in the Program Code I've made, the Earth's orbit follows a perfect circle. In reality, it follows a slightly elliptical path as it travels around the Sun. (Newton's First Law states that all orbits follow paths that are conical sections; therefore, an orbit can be circular, elliptical, parabolic, or hyperbolic.) For simplicity, my program places the asteroids in elliptical orbits such as those followed by asteroids 1,577 Icarus, 1,862 Apollo and 1,221 Amor.

To increase the accuracy of your simulation, you can use the fact that the ratio of the Earth's diameter to the length of its orbital circumference around the Sun is approximately $8,000/(2\pi 93,000,000)$ or $1/73,000$.

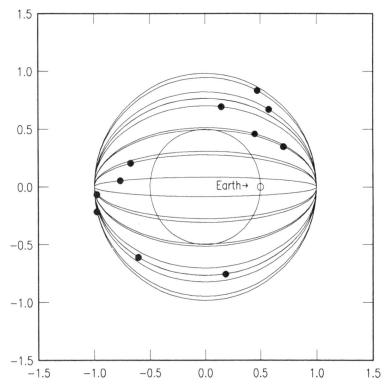

Computer-generated orbits. How many intersect with Earth?

Bernard Peter Gore, author of *The Death of Nemesis*, predicts less than one impact every 100,000 years assuming 10,000 bodies in Earth-crossing orbits (also assuming their orbits are all within 5 degrees of the ecliptic). Although Gore believes a kilometer-sized object (especially a comet) would not cause a global catastrophe, an object several hundred kilometers in diameter would devastate the Earth. Such impacts should occur every million or 10 million years.

Secret Impacts Revealed

For 20 years, military satellites have been observing huge meteoroids slam into the Earth's atmosphere, but until recently the U.S. Department of Defense kept the information classified. The figure on p. 131 shows the "secret" locations of 136 meteoric airbursts occurring near the Earth from 1975 to 1992—information obtained by satellites of the U.S. Air Force Space Command. Although the Air Force doesn't disclose the exact nature of its surveillance system, the data probably comes from early-warning satellites of the Defense Support Program.

Just how real a threat are impacts from outer space? Chapter 21 discusses this in greater detail, but consider the following sobering statistics:

- On average, every year a fragment of asteroid or comet blows up somewhere in our atmosphere with an energy equivalent to 20,000 tons of TNT.
- These objects are 10 meters across, weigh 1000 tons, and travel at 20 kilometers a second.

- Recently, the 36-inch Spacewatch telescope in Arizona observed numerous house-size objects hurtling through space perilously close to Earth.

- According to D. Rabinowitz in the April 10, 1993, *Astrophysical Journal*, the Earth receives more encounters with outer space objects than previously predicted: a 20-kiloton blast occurs roughly every month, and the Earth receives hundreds of kiloton-yield jolts annually.

- On April 15, 1988, a U.S. satellite recorded the blast of a behemoth fireball above Indonesia. For an instant, the explosion was brighter than the Sun. Military experts conclude that the blast was from a large meteoroid exploding with the energy of 5000 tons of TNT.

- On August 3, 1963, an asteroid fell near Antarctica with an explosive punch of a half million tons of TNT.

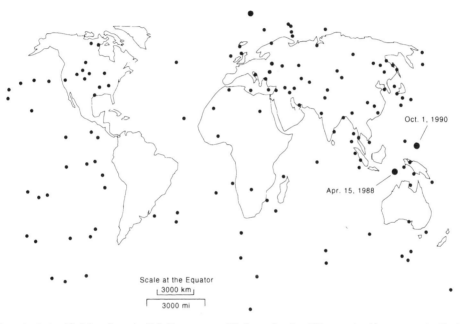

Recently declassified data from the U.S. Department of Defense showing 136 meteoric airbursts near the Earth from 1975 to 1992. (After Beatty, 1994.)

The Millerites

The earth and all the works therein
Dissolve, by ranging flames destroyed;
While we survey the awful scene,
And mount above the fiery void.
—Millerite Hymn, 1843

We have so far dealt with astronomical predictions of the Apocalypse. However, throughout history many religious groups have found the concept of the Apocalypse irresistible.

The most interesting religious group concerned with the end of the world were the

Millerites who based their teachings upon the studies of William Miller, a fundamentalist Protestant who predicted that the world would end in 1843. He based his prediction on the following biblical passage in Daniel 8:14:

> And he said unto me, Unto two thousand and three hundred days; then shall the sanctuary be cleansed.

Since the prophesy was made in about 457 B.C., the end of the 2300 year period would come in 1843. (Miller believed that the word "days" in the biblical passage should be translated as "years.") In 1842, Millerite leaders passed the following resolution:

> Resolved: that in the opinion of this conference, there are most serious and important reasons for believing that God has revealed the time of the end of the world, and that that time is 1843.

The Millerites deeply believed in the end of the world. Many Millerites actually attempted to fly bodily to heaven. Some followers donned wings, climbed trees, and prayed to the Lord to lift them up. Inevitably several followers broke their arms when they jumped from the trees.

Today, the Seventh Day Adventists and Jehovah's Witnesses believe that end of the world is near. Most Protestant groups share the basic idea that the world could end at any moment.

Behold the Terrible Ragnarök

> *The sun turns black, earth sinks in the sea, the hot stars fall from the sky, and fire leaps high above heaven itself.*
> —"The Edda" (Icelandic poem about the pre-Christian Norsemen)

Aside from various Christian sects, the pre-Christian Norsemen were fascinated by the end of the world. Christians believe that after Doomsday the soul of man would be immortal, and they believe God is immortal. To the Norsemen, Doomsday (called Ragnarök) means the death of everything, everyone, including the gods. One Icelandic poem describes Doomsday for the Norsemen: "The gods are doomed, and the end is death."

The Ragnarök is preceded by Fimbulvetr, a horrible winter lasting for three years. After this, the real action begins:

- The wolf Fenrir, whose jaws stretch from heaven to Earth, breaks his chains.
- The Midgars Snake rises from the sea and spits poison.
- Naflfar, the Ship of the Dead, made of the uncut nails of dead men, breaks from her dock in Hell.
- Surt leads the fire giants from Muspell. They approach Asgard (the home of the gods), and Bifrost (a rainbow bridge) cracks.

And this is just the beginning. Suffice it to say, all hell breaks loose. Thor, the protector of humankind, is killed. The human race is dumped off the Earth. Surt, the fire giant, sets the earth and heaven on fire, and finally the entire Earth sinks beneath the sea.

Of course, the Norsemen were not the only group to give explicit and dramatic information pertaining to Doomsday. St. Paul in the Bible predicts:

The Lord himself will descend from heaven with a cry of command, with the archangel's call, and with the sound of the trumpet of God. The dead in Christ will rise first; then we who remain alive shall be caught up together with them in the clouds to meet the Lord in the air; and so we shall always be with the Lord.

We shall not all sleep, but we shall all be changed, in a moment in the wink of an eye, at the last trumpet. For the trumpet will sound and the dead will be raised imperishable, and we shall be changed. (I Thessalonians 4:16–17)

Daniel Cohen, in his wonderful book *Waiting for the Apocalypse*, notes that Paul speaks of "we who remain alive." This suggests that St. Paul believed that the Apocalypse was to take place within his own lifetime and the lifetime of the majority of his listeners.

In the past, numerology was often used to decipher the meaning of the Book of Revelation without much controversy; however, sometimes numerology and astrology went a little too far. For example, Cocco d'Ascoti was burned at the stake in 1327 for trying to cast Christ's horoscope.

Gnostics, Montanists, and Shiptonists

Montanism, which had for years been a powerful alternative to orthodox Christianity, went into decline, because it could not adapt to a world that stubbornly refused to end.
—Daniel Cohen, *Waiting for the Apocalypse*

Every meteor in the sky seen at Jerusalem brought the whole Christian population into the streets to weep and pray.
—Daniel Cohen, *Waiting for the Apocalypse*

There are several other religions devoted to the Apocalypse or eschatology. *Gnostics* in the 1st century A.D believed that the world had already ended, and that life on Earth was meaningless and unimportant.

Montanists, named after their Christian leader named Montanus, in the 2nd century believed that a heavenly city would land at a place on Earth called Ardabau—a signal for the Second Coming. In A.D. 198 some Montanists actually claimed to see a walled city floating in the sky over Judea for 40 consecutive mornings. (See Chapter 3 regarding of the number 40 in many religions.) Unfortunately, Montanists actively sought martyrdom, and the Romans were only too happy to oblige them.

Shiptonists in the 15th century followed the English prophetess Mother Shipton who predicted that the world would end in 1881. This small waif of a woman had a reputation for startlingly accurate predictions. Unfortunately for the Shiptonists, it turned out that Mother Shipton was a fabrication of an enterprising publisher who wrote her prophesies after events happened but claimed that they came in advance. The publisher sold a very successful book on the imaginary Mother Shipton.

Modern day apocalyptic sects seem to grow every year. One interesting example is the True Light Church of Christ, the leaders of which taught their 450 followers in North and South Carolina that the world would end in 1970. On January 2, 1971, *The New York Times* interviewed a member of the sect who said he had not yet decided whether to reopen the upholstery shop that he closed a year previous in preparation for the end.

So far these prophets who have predicted the end of the world have had a perfect score—they have always been wrong.

In more traditional Christian sects, there are many who interpret words in the Book of Revelation to mean the coming of an Apocalypse. Apocalyptic prophetic literature began to emerge during the Babylonian exile, the time when the Jews were taken to Babylon (Iraq) as slaves. Before this point in time, there was little in the way of apocalyptic literature. (It was thought that God would restore Israel.) During the Greek occupation, this theme was revived much more strongly in Daniel and the book of Enoch (pseudepigrapha). One of the emergent themes was individual justice (if the evil foreign empires are not punished in this life, it must be that they will find their punishment in the next life). This concept was well entrenched by the time Jesus was active.

One of the current questions of theology is whether the historical Jesus believed in the eschaton as something that comes at the end of all time, in a miraculous event by God, or whether it is something made manifest in daily life. Some feel that the current Church is living that question with the same ambiguity reflected in the Gospels. The coming Christian millennium is a focus for some who expect an eschaton.

The Lord is merciful. He maketh me lie down in green pastures. The problem is, I can't get up.

—Woody Allen, *Without Feathers*

CHAPTER 11

Stonehenge

Ancient civilizations built the foundations of modern science [and math] on the study of cyclic events, usually astronomical ones. Sunrise and sunset, tides, the phases of the Moon, the annual march of the seasons, even the 18-year repetition in the pattern of solar eclipses, were recognized more than 2000 years ago.

—Paul R. Weissman, *Sky & Telescope*, 1990

When the Cimetière des Innocens at Paris was removed in 1786–1787, great masses of adipocere[1] were found where the coffins containing the dead bodies had been placed very closely together.

—Adipocere, *Encyclopedia Britannica*, 11th Ed.

On the broad Downs, under the gray sky, nothing but Stonehenge, and all round, wild thyme, daisy, meadowsweet, and the carpeting grass.

—Ralph Waldo Emerson

"Sir, where are we?"

Mr. Plex is stepping in a bed of stonecrop: showy flowers of white, yellow, pink, and red. A vine of pink orphine with blue-green leaves trails along his abdomen.

"Help me up," Theano says, extending you her arm.

Slowly rising, you stretch your sore legs, and pull her up.

Theano takes a deep breath, smiles, and slowly turns to take in her surroundings.

The three of you are standing in the center of a circular structure of large upright stones.

You smile back at her as you run your hand through her long brown hair. "Enjoy the new setting?"

"You behave," she says wickedly.

Mr. Plex sneezes. Perhaps it is the rich scent of honeysuckle.

"Sir, what month is it?"

"December."

"Then why are flowers blooming?"

"Not sure. Maybe the transfinites are playing with us?"

Mr. Plex removes a flower from his abdomen. "Or maybe our presence in ancient Greece has altered the fabric of time. You know, the butterfly effect in chaos theory. The flapping of the wings of a butterfly can change the weather patterns."

You gently touch some golden stonecrop forming a mosslike mat on the huge rocks. "We're eight miles north of Salisbury, England."

"Sir, you took us to Stonehenge?"

Earliest known perspective drawing of Stonehenge from a Dutch manuscript, 1574. (Early drawings have significant inaccuracies.)

You nod. "The ring of stones is about 350 feet in diameter." You motion toward a gap in the circle of stones to the northeast. "That block of sandstone weighs 35 tons. And if my information is correct, we're surrounded by an outer circular ditch and a ring of 56 pits."

Theano lets out a whoosh of air. "Who built this? What's it for?"

You start walking closer to the ring of holes as you turn back to Theano. "Probably a place of worship, but we don't know much about the religion. Some of the stones are aligned on special areas of the horizon where the sun and moon rise at important times of the year." You pause. "My opinion is that Stonehenge was a temple for sky worship. Most scientists believe it flourished as a temple or meeting place for more than one thousand years."

Mr. Plex's legs sink slightly into the sod. "Sir, why speculate? Why don't we just go back in time to whenever people worshiped here and find out?"

"I don't want to risk another disruption of the distant past."

Mr. Plex nods. "We did cause quite a stir in ancient Greece."

You nod. "Maybe the transfinites will leave us alone if we stay away from people."

Theano slips her hand through the crook of your arm. "Which way do we head?" she asks, moistening her lips.

You shift the statue of Dionysus from your left to right hand.

Theano wrinkles her nose. "Do we have to keep that?"

You shrug. "Follow me."

You lead Mr. Plex and Theano to a ring of pits, stoop down, and gaze into a dark hole. "These 56 pits are known, after their discoverer, as Aubrey holes."

Mr. Plex's eyes seem to glaze over. "56. A very interesting number. It's the 6th tetrahedral number. The sequence is: 1, 4, 10, 20, 35, 56, 84, 120 … with a generating formula $(1/6)n(n + 1)(n + 2)$."

You put up your hand. "That's all very well and good Mr. Plex, but I don't think it has much to do with Stonehenge."

Mr. Plex ignores you and looks at Theano. "Consider an example using cannonballs. The number of balls in each layer is, starting from the top of the pile, 1, 3, 6, 10, 15, … which forms a sequence of triangular numbers because each level is shaped like a triangle. Tetrahedral numbers can be thought of as sums of the triangular numbers."

You clap your hands together. "Mr Plex!"

Stonehenge drawing from Camden's Britannia, 1586.

"One minute sir." He pauses. "We can extend the idea into higher dimensions, into hyperspace." He says the word "hyperspace" with a trace of awe. "In 4-dimensional space, the piles of tetrahedral numbers can themselves be piled up into 4-dimensional hypertetrahedral numbers: 1, 5, 15, 35, 70.... We can form these numbers from the general formula: $(1/24)n(n + 1)(n + 2)(n + 3)$."

You decide that the best way to stop Mr. Plex is by whispering just a few words. You motion to one of the Aubrey holes and say, "There are legends of treasure inside ..."

Theano quickly reaches inside one of the pits.

You pull her hand out, crouch low, and begin to place your hand in the pit's murky interior. "Wait. Let me make sure it's safe."

Drawing by Dr. Charleton, 1663.

"The hell with that," Theano says. "Let me see what's in there."

Now Theano pulls *your* hand out of the hole, reaches inside, and brings out some grey, gritty powder.

Your left eyebrow raises. "Cremated burial remains and bones."

Theano's eyes open wide with horror as she throws dried adipocere[1] at you.

You turn to Mr. Plex. "Even though two-thirds of the excavated Aubrey holes contain cremated human bones, the purpose of the holes is not necessarily sepulchral. In fact, you can find other cremated burial remains in the bank and in the silting around the ditches."

Theano takes a deep breath. "Where are all the other holes?"

You glance at tiny bone fragments in the hole and then look at Theano. "Holes 33 to 54 have never been excavated. Their locations are known by bosing—sounding the ground with a large hammer and listening for soft spots."

"Sir, if the holes aren't some kind of burial pits, what are they for?"

"Some scientists say the holes, as well as the stones, function as a computer or counting device."

"You mean Stonehenge is some kind of Stone Age computer?"

You nod. "Four station stones mark the winter and summer sun. They also mark the moon through all turning points of the 56-year moon cycle. The critical sunrises, moon-rises, sunsets, and moonsets are framed by the archways of the *sarsen*—the large boulder—circle as seen through the trilithons."

Mr. Plex ambles closer to some large stones. "Trilithon?"

You motion to the structures consisting of three large stones: two upright and one resting upon them as a lintel:

"That's interesting," says Theano, "but how does that make it a computer?"

You nod. "I'm getting to that."

Mr. Plex comes closer. "How old is this place?"

"Humans were constructing and altering Stonehenge for over five centuries—about 25 generations." You motion all around you. "All of this was begun in the same millennium as the Great Pyramid of Giza, a few centuries before Abraham dwelt in Canaan!" You pause. "The Aubrey holes date back to 1800 B.C." You point to tall blue rocks. "Those pillars, called bluestones, were brought here in the 17th century B.C."

You walk to a circle of 30 upright stones capped by a continuous ring of stone lintels enclosing a horseshoe formation of five trilithons. From your history books, you know that the uprights are 30 feet long and weigh 50 tons.

Theano tugs at your hand. "Why did you call it a computer?"

"Stonehenge was built with its central axis pointing to the midsummer sunrise. It was a giant calculator for predicting solar and lunar eclipses. Several researchers say that Stonehenge builders had a knowledge of Pythagorean geometry 3000 years or more before the Greeks." You pause. "Look at this. The archways and stones point to the rising and setting of the sun and moon at critical dates of the year. For example, the Moon rises farthest to the north when it appears over that stone as seen from the center of Stonehenge."

Mr. Plex's abdomen is pulsating. "It could predict eclipses? Imagine what fear a Stonehenge reader could cause if he could predict and seemingly control eclipses. Maybe a priest would strike terror in his people."

"A Professor G. S. Hawkins in the 1960s showed how Stonehenge might be used as a computer to predict the time when eclipses of the Sun and Moon were due. He ingeniously related the 56 Aubrey holes with a 56-year eclipse cycle. Hawkins also suggested that cremations were performed in a particular Aubrey hole during the course of the year as a kind of bookkeeping system!"

You take a deep breath, trying to recall Hawkins' words. "Here is a typical recipe Hawkins gave for using this Stone Age monument as a digital computer:

Take three white stones, and set them at Aubrey holes 56, 38, and 19. Take three black stones and set them at holes 47, 28, and 10. Shift each stone one place around the circle every year, say at the winter or summer solstices. This simple operation will predict accurately every important lunar event for hundreds of years.

"It's all Greek to me," Theano says tossing up her hands.

You smile. Evidently Theano is catching on to the idioms she hears from you.

"The point is that the monument could certainly form a reliable calendar for predicting the seasons. It could also signal the spooky periods for an eclipse of the Sun or Moon."

"Sir, humans have always held eclipses in awe—sudden darkness at noon as the Moon blots out the Sun, the orange glow of the Moon in the Earth's shadow."

You nod. "Stonehenge wasn't the only great stone monument. Humans built monuments like Stonehenge all over Europe, until about 1000 B.C. People may have continued to congregate and worship at the stone rings, and its possible that a Druidic priesthood used them as temples."

"Druids?" Theano asks.

A shiver seems to run up Mr. Plex's spine. "Sir, they were disgusting."

"Ancient Celts. Priests. Sorcerers. They believed in reincarnation." You pause. "They weren't the nicest guys in the world. The Druids sacrificed humans to help others who were sick or in danger of death in battle. Huge wickerwork sculptures were filled with living men and then burned. Although the Druids preferred to burn criminals, innocent people were burned if criminals were in short supply."

Theano shudders. "What stone is this?" she asks as you walk toward the north.

"Ah, the slaughter stone. See the channels on the rock?"

She nods.

"Through the channels ran the warm blood of Druid victims."

Theano's eyes open wide. "Really?"

You wave your hand. "Just kidding. Most likely the channels are natural to the rock. Even so, the name 'slaughter stone' is still used."

Mr. Plex, in his excitement, suddenly bangs his massive body into one of the trilithons. It falls into its neighbor and, one by one, in a massive domino effect, all the remaining trilithons start crashing to the ground. The reverberations echo through the hillsides as dust and debris ascend like a phoenix rising from the ashes.

"Mon Dieu!" Mr. Plex screams looking at the chaos around him. "What do we do now?"

You are quiet, calm, always in control. "Can you put the stones back the way they were?"

Mr. Plex eyes the massive stones. "I can try."

Theano and you watch as Mr. Plex drags the stones into position, occasionally uttering strange grunts signifying his awesome effort. Even the most proficient human sopranos or basso profundos are limited in the range, duration, and timbre of their voice due to the physical constraints of their vocal apparatus. Mr. Plex, a scolex, has no such limits. His grunts and groans are like a dozen bassoons filled with whipped cream.

After several hours of abdomen-breaking work, Mr. Plex repositions the last of the trilithons. "I think I got *most* of it right," he says.

You nod. "We better get out of here. The transfinites aren't going to like this one bit."

Stonehenge would be functioning quietly and unnoticed as it had for the past 4000 years,
a lonely monument to the scientific ability of prehistoric man.

—Gerald Hawkins, *Beyond Stonehenge*

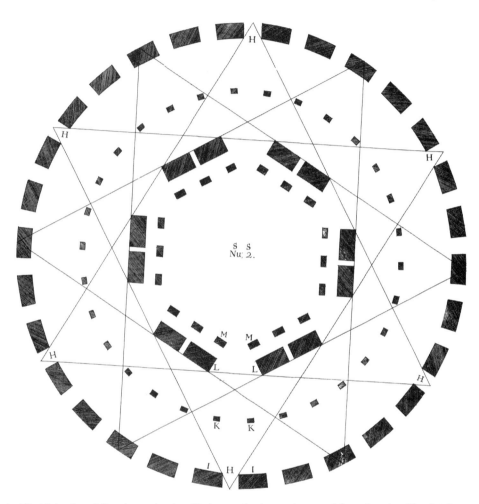

Architect Inigo Jones' Stonehenge drawing. He incorrectly shows a hexagonal figure based on 12-pointed star. What is the length of the side of the internal 12-sided isogon, if the side of the equilateral triangle is 1 unit? (Most mathematicians to whom I spoke suggest that the answer is $\frac{1}{2} - \sqrt{3}/6$. The edges of the interior regular 12-gon are of length $2\sqrt{3}/3 - 1$.

All information on Stonehenge is based on current scientific documents (see References). The cremated remains of human bodies have been found in the Aubrey holes. Various authors have speculated as to the possible astronomical significance of the placement of stones within Stonehenge.

Druids did exist, although most of what we know about these priests and learned class of Celtic people is derived from secondhand Greek, Roman, and medieval Irish sources. Caesar referred to Druids as priests, religious ministers, or teachers. In native Irish and

Welsh legends, the Druids are magicians and sorcerers. Like the Pythagoreans, the Druids' principle doctrine was that the soul was immortal and passed at death from one person into another. Ironically, the Druids' sacrificial rites were conducted with living people placed in huge wicker cages and burned. I find this difficult to understand considering that a Druid individual must have realized that this practice insured an individual would eventually be burned alive in a later incarnation.

How did the Druids get their name? Because the oak was sacred to the Druids (as was the mistletoe), the term "Druid" may be derived from the Celtic word for oak tree, *daur*. Incidentally, there was also a lower ranking Druidic priesthood, the *vates*, who were prophets and bards.

The legends of the Druids were preserved in Irish epics for several centuries after the Christianization of Ireland. There are faint undercurrents in Welsh folklore. The doctrines of the Druids seem to have disappeared from Europe in the early Middle Ages. In the 17th century, scholars believed that the Druids had constructed Stonehenge, but now we believe Stonehenge is of earlier origin.

Famous Stonehenge explorer Inigo Jones (1573–1652), architect to James I and Charles I. (See his drawing on page 141.)

Stonehenge explorer Walter Charleton, M.D., physician to Charles II.
(See his drawing on page 138.)

Look round the world: contemplate the whole and every part of it; you will find it to be
nothing but one great machine, subdivided into an infinite number of lesser machines ...
We are led to infer that the Author of Nature is somewhat similar to the mind of man.
                                                                                    —David Hume

CHAPTER 12

Urantia and 5,342,482,337,666

The Christians know that the mathematical principles according to which the corporeal world was to be created are co-eternal with God. Geometry has supplied God with the models for the creation of the world. Within the image of God it has passed into man, and was certainly not received within through the eyes.

—Johannes Kepler (1571–1630)

If we find the answer to [why we and the universe exist], it would be the ultimate triumph of human reason—for then we would truly know the mind of God.

—Stephen Hawking, *A Brief History of Time*

"Mr. Plex, what's the religious significance of the number 5,342,482,337,666?"

Mr. Plex taps upon a trilithon and looks up. The sky is the color of slate now as daylight fades. The dark flowers around Stonehenge seem to come to life, and the moss on the stones takes on a ghostly lunar glow.

Mr. Plex finally turns to you. "I'm not sure."

"How about the number 611,121?"

"Sir, did you say *religious* significance?"

You nod. "In the Urantia religion, a 20th century sect, these numbers have an almost divine quality. Urantia is their name for Earth." (You pronounce it you-RAN-sha.) "According to the sect, headquartered in Chicago, we live on the 606th planet in a system called Satania, which includes 619 flawed but evolving worlds. Urantia's grand universe number is 5,342,482,337,666." You pause. "Satania has its headquarters at Jerusem in the constellation of Norlatiadek, part of the evolving universe of Nebadon. Nebadon in turn belongs to a super-universe called Orvonton. Orvonton and six other super-universes revolve around the central universe of Havona, the dwelling place of God."

Theano is reclining beside an Aubrey hole, coiled in the flickering shadows. "You've told me about Christianity," she says. "How does Urantia compare?"

"Urantia followers rely on a 2097-page holy book. They say it contains the Earth's fifth revelation from God, superior to mainstream Christianity. They think it will transform the world." You pause. "In their book, Lucifer, who rebelled against his superiors, is now the deposed sovereign of Satania, named after Satan, his first lieutenant."

"Sir, how did they come up with 5,342,482,337,666?"

"I don't think anyone knows. Nor does the Urantia movement disclose who wrote *The Urantia Book*, which mentions the number. They do believe that human minds are created at birth, but the soul does not develop until about age six. When we die, our souls survive. Jesus Christ is number 611,121 among more than 700,000 Creator Sons."

Theano leans closer to you. "Pythagoras would have liked this religion. A numerologist's dream."

"Sir, do we know if 611,121 has any particularly interesting numerical properties?"

You nod. "Theano, could you hand Mr. Plex the notebook computer?"

"OK." She cautiously reaches into an Aubrey hole. After a few seconds she brings out a computer coated with the ashes of burned bones. She hands it to Mr. Plex who begins to type, slowly at first, but then with exponentially increasing speed.

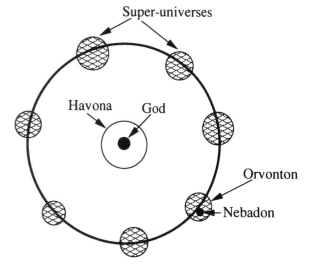

The universe according to the Urantia movement. Pictured are seven super-universes circling around the central universe of Havona, dwelling place of God. We live somewhere in the evolving universe of Nebadon within the super-universe called Orvonton.

"Sir, I'll write a program to find the prime factors of each of the numbers. Maybe they'll have some clues."

After a few seconds, a speech synthesizer chip in the computer says:

$$611121 = 3 \times 7 \times 29101$$
$$5342482337666 = 2 \times 829 \times 1361 \times 2367557$$

"Sir, both numbers don't seem to have very many prime factors. How many prime factors would you expect to find for numbers this big?"

You scratch your head. "I'm not sure, but let me tell you about some amazing numerical properties of these numbers. For example, consider the digits of 611,121 one at a time:

$$6 + 1 + 1 + 1 + 2 + 1 = 6 \times 1 \times 1 \times 1 \times 2 \times 1 = 12$$

And 12 plays a major role in many religions. For example, there are 12 signs of the zodiac. Christ chose 12 apostles. In the Revelation to John, the heavenly Jerusalem has 12 gates. There are even 12 tones in our modern 12-tone musical scale."

"Mon Dieu! That's incredible."

You turn to Theano. Is that a trace of admiration you see in her eyes? If so, you plan to turn her *admiration* into *adoration* with your next revelation.

Before you have a chance to speak, Mr. Plex shouts out, "Mon Dieu! The factor 1361 in the grand universe number—it's the exact year during the papacy of Gregory XI when he decided to move from Avignon to Rome, thus ending the Babylonian captivity of the Church."

You stand up and look from Theano to Mr. Plex. "611,121 appears in a listing of decimal representation periods for $1/n$ for $n = 1,2,3, \ldots$:

```
1,1,1,1,1,1, 6,1,1,1,2,1,
6,6,1,1,16,1,18,1,6,2,22,1,1,6,3,6,28,1,
15,1,2,16,6,1,3,18,6,1,5,6,21,2,1,22,46,1,42,1,16,6,13,3,2,6,
18,28,58,1,60,15,6,1,6
```

In fact, 611,121 is the very first string of digits, starting with a number other than 1."

Theano screams out, "Mon Dieu!" in an imitation of Mr. Plex. Then she pauses. "But what the hell does that mean?"

You grin. "Examining the decimal representations of fractions, 1/1, 1/2, 1/3 ,1/4, 1/5, 1/6, …, it's not until 1/7 do we get a 'period' of 6 since $1/7 = 0.142857142857$, which repeats every 6 digits. The next 2 in the sequence indicates that $1/11 = 0.0909090$ repeats every 2 digits."

You gaze into the smoldering depths of Theano's eyes. Your voice is deep. "Could the Urantia movement have known about the unusual characteristics of their number for Christ?" Then you whisper. "What other startling numerical characteristics do they have?"

Before Theano and Mr. Plex have a chance to think, you see bats alighting upon some of the largest stones at Stonehenge. Their black wings quiver on slender bodies. Whenever the moonlight reflects off their eyes, they glow like little neon lights. A few of the bats seem to have developed a large fold of skin between their arms and legs. When the animals jump off a trilithon, the skin balloons upward like a parachute to slow down their rate of descent.

A few bats circle overhead ever-so-slowly, as if suspended in space. Soon a few more bats join them. As the bats come closer, you catch a glimpse of their sunken faces. Their crimson eyes seem to radiate hatred and torment.

In the distance are dull pulses of thunder. You look at Theano and see goosebumps on her arms.

Stonehenge is beginning to make the blood slide through your veins like cold needles.

THE SCIENCE BEHIND THE SCIENCE FICTION

No holy Bible offered to the Western world in the past few centuries is thicker, heavier, or stranger than The Urantia Book. *This 2,097-page, 4.3 lb. volume purports to be written entirely by superhumans and channeled through an unknown earthling.*
—Martin Gardner, *On the Wild Side*

A Urantia movement is headquartered in Chicago. The members of the movement believe the revelation in *The Urantia Book* is superior to mainline Christianity and destined to change the world.[1]

Now in its tenth printing, *The Urantia Book* was first published in 1955 at the suggestion of surgeon William Sadler, a one-time ordained Seventh-Day Adventist minister. The Urantia movement does not disclose who wrote *The Urantia Book*, and Dr. Sadler has written, "We are not at liberty to tell you even the little we know about the technique of the production of the Urantia papers …" Martin Gardner notes that *The Urantia Book* is a mixture of Seventh-Day Adventist doctrines and Adventist heresies. One major deviation from conservative Adventism is the book's full acceptance of evolution and the great antiquity of the earth.

The Urantia Book describes in great detail the structure of the universe and God's administration. Countless subdeities and angels serve God. "Technical Advisors" include: Supernaphim, Seconaphim, Tertiaphim, and Omniaphim. The "Master Physical Controllers" include: Power Directors, Mechanical Controllers, Energy Transformers, Frandalanks, Chronoldeks, and John the Baptist. *The Urantia Book* also speaks of a finite deity who is evolving into the Supreme Being of all evolving universes.

The first two humans on Urantia are not Adam and Eve, but rather the twins Andon and Fonta, children of beasts. Followers of the Urantia movement do not believe in the concept of original sin.

As you told Mr. Plex and Theano, the numbers 611,121 and 5,342,482,337,666 play significant roles in the Urantia movement. Here are some additional numerical observations:

$$611{,}121 = 2^2 + 34^2 + 781^2$$

$$611{,}121 = 15^3 + 52^3 + 55^3 + 67^3$$

$$611{,}121 = (1! + 3!) \times ((1^3 + 2^3 + 3^3 + 4^3) \times (1! + 2! + 3! + 4! + 5! + 6!) + (1! + 2!))$$

$$611{,}121 = \sum_{i=1}^{i+405} i + \sum_{i=1}^{i=1028} i$$

$$611{,}121 = \sum_{i=1}^{i+579} i + \sum_{i=1}^{i=941} i$$

$$611{,}121 = \sum_{i=1}^{i+1} i + \sum_{i=1}^{i=10} i + \sum_{i=1}^{i=1105} i$$

$$611{,}121 = \sum_{i=1}^{i+1} i + \sum_{i=1}^{i=115} i + \sum_{i=1}^{i=1099} i$$

$$611{,}121 = \sum_{i=1}^{i+1} i + \sum_{i=1}^{i=415} i + \sum_{i=1}^{i=1024} i$$

$$611{,}121 = \sum_{i=1}^{i+1} i + \sum_{i=1}^{i=529} i + \sum_{i=1}^{i=970} i$$

Here the Greek letter sigma, Σ, represents the summation sign in mathematics, and is used to denote the sum of a series, so, for example, $\sum a_r = a_1 + a_2 + a_3 + \ldots + a_r + \ldots + a_n$.

Perry Bowker from Toronto, Canada, sent me additional information of potential interest to numerologist readers:

> Let us consider the Urantia number 5,342,482,337,666. If you strip away the last 666, then 5,342,482,337 is close to the human population of the Earth right about now, and 666 traditionally represents old Scratch himself, so this is code for "the world has gone to hell."

A French mathematician on the Internet comments:

> Regarding 611,121, I seem to recall that the first examples of non-isometric isospectral surfaces of curvature -1 are close to this genus number. They were discovered by M. Ville in the '60s.

(The mathematician is referring to a branch of mathematics called differential geometry which deals with three-dimensional curves and surfaces.)

I thank the Urantia Number Cartel, a group of friends who helped research the numerical properties of 611,121 and 5,342,482,337,666. The cartel convened in November, 1994, and included: J. Kleid, S. Frye, D. Shippee, and R. Fung. I made the startling discovery of the relation between 611,121 and the periodicity of 1/n sequences with the aid of Neil Sloane's "On-Line Encyclopedia of Integer Sequences," AT&T Bell Labs, Murray Hill, New Jersey.

The program code in the Appendix shows you how to determine the prime factors of the Urantia numbers.

數 數 數 數

In this chapter, as in others, we've seen the prominent role that integers play in religion. However, the Urantia religion appears to attribute significance to multidigit numbers (e.g., 5,342,482,337,666), more so than any other religion with which I am familiar. It is clear that since the time of Pythagoras humans have attributed to integers certain properties and features usually attributed only to God. According to Philip Davis and Reuben Hersh, authors of *Descartes' Dream*, the religious philosophies of Judaism, Christianity, and Islam suggest that God is independent of the world, and he will persist even when the world has come to an end. Many mathematicians also believe that mathematics is independent of the world, that it exists prior to and apart from the world, and when the world ends, mathematics will still exist. Moreover, Davis and Hersh note:

> Insofar as Platonic mathematics is a rival of an eternal, all knowing, omnipotent God, it is not an accident that many people turn to mathematics, consciously or unconsciously, as a substitute for religion. There is a strong craving for permanence, for certainty in a chaotic world, and many people prefer to look for it within a mathematical or scientific rather than a religious context. They are, perhaps, not aware that underlying both mathematics and religion there must be a foundation of faith which the individual must himself supply.

CHAPTER 13

Fractals and God

The belief that the underlying order of the world can be expressed in mathematical form lies at the very heart of science. So deep does this belief run that a branch of science is considered not to be properly understood until it can be cast in mathematics.

— Paul Davies, *The Mind of God*, 1992

A fractal pattern, Mandelbrot's—
a mapping of the complex kind—
expanded like a blooming Rose—
a flower of the Mind.
But even Beauty couldn't hide—
beyond the petals' filigrees—
the Hand that sprays Insecticide on all of our Infinities.
— Keith Allen Daniels, "Number of the Beast for Emily Dickinson and the Em Dash"

God is a mathematician.
— Astronomer James Jean

"I wish you'd get rid of that thing," Theano says.

The statue of Dionysus, your memento from ancient Greece, now has a large head and mouth, small eyes, and a bumpy skin covered with wartlike lumps and fleshly flaps. Its motionless eyes blend with the color of its surroundings like a chameleon. Dionysus' head reminds you of a stonefish, a species of venomous marine fish of the genus *Synanceja*.

You nod. "You're probably right. Still, I'm curious to see what happens to it."

Theano points to small insects that alight on the mutant Dionysus. "What are they?" she asks, gazing at their long antennae.

You stoop down. "Stoneflies, order Pecoptera. Weak chewing mouth parts. Membranous wings. Each female produces 6000 eggs. The babies have external gills. The hindwings fold like a fan."

Mr. Plex squats down to get a closer look. "Sir, I like their vein patterns."

You nod. "Fractals."

"Fractals?" Theano says.

"Intricate patterns repeating at different size scales. Branches within branches. Mathematicians call such structures *self-similar* because similar structures are present at different magnifications. My favorite self-similar fractals are the Julia sets."

Mr. Plex jumps up. "Sir, can we program them on a computer?"

You reach into an Aubrey pit and pull out a notebook computer. Unfortunately, the charred dust of ancient bones coats the display, and a mandible clings to the keyboard. You toss the notebook computer to Mr. Plex. The mandible hits him in the neck.

"Let's make some fractal Julia sets," you say. "They're based on mathematical feedback loops. You square a complex number, add a constant, and continue to repeat the process."

"Complex numbers?" Theano asks.

"Complex numbers have a real and imaginary part. Modern mathematicians deal with them all the time."

Mr. Plex wipes the bone remnants from the display and anxiously begins to type on the computer. "Ah, I recall learning about them. Any portion of the Julia set may be magnified again and again, and each new magnification brings forth new riches and delights."

You nod. "The sets are generated by successive applications of the rule, or mapping, $z \rightarrow z^2 + c$, where c is a constant." You pause. "Another related fractal is the Mandelbrot set. To create this bushy shape, you start with $z = 0$, and continually iterate (repeat) $z \rightarrow z^2 + c$. The Mandelbrot set is a sort of map for all the Julia sets because it tells us about the behavior of different c values under iteration. If points c are small and start inside the Mandelbrot set, they can't escape. In fact, points in the main cardioid swirl like a whirlpool as they drain into a fixed point. Points outside the prickly shape fly off to infinity."

"Sir, the edge separating trapped points and exploding points is quite intricately shaped."

"Quite right. The edge is fractal—it's rough with incredible detail." You press a few buttons on the computer, and beautiful colorful images appear. They look like glistening jewels in the early evening light. "A point that starts its life in the black region is forever confined in a prison with intricate walls, trapped, never knowing the freedom of its infinitely near brethren—brethren born outside the crinkly edge exploring far and wide, eventually exploding outward to meet their God, infinity." You pause to assess the effect of your poetic words on Mr. Plex and Theano.

They are speechless.

The sky is the color of ash now, and the daylight is fading. The air smells of ozone and roses.

You point to little black specks within the spirals on the screen. "That's a close-up of the Mandelbrot set, a magnification of a tiny part of the original. Each of these bumpy-warty bits would look just like the original bushy shape if we magnified them. Currently we're looking at a section magnified 1000 times, as if we were studying geometry with a microscope."

The glowing spirals, with the nodes and turns, enchant you in the darkness of Stonehenge. There are white and green spirals upon aquamarine spirals in an endless intricate cascade. Here and there is a touch of crimson. The colors are so vivid as to nearly make you weep. The computer is like a microscope opening a portal to a vast, unexplored and unpredictable universe. Who would have thought that such beauty could be hidden within what was initially so lumpy looking? How could such a simple formula produce an inexhaustible reservoir of magnificent shapes and forms?

Your face had always been sharp-featured and deeply furrowed, but right now when you catch a reflection of yourself in the glass of the monitor, it seems more relaxed. It must be the fresh air, Stonehenge, and the invigorating day.

You press a button on the keyboard. "Let's magnify this spot a bit more." As you continue zooming in, one of the spirals seems to unfurl and blossom, exploding outward to angular infinity. It reminds you of ocean waves. You can almost smell the surf. The space is so vast. "Freedom" is the only word that comes to your mind.

"Sir, do you think that mathematics and fractals are an invention of sentient beings like ourselves, or do they have an independent existence?"

"Good question, Mr. Plex. Some mathematicians believe fractals and mathematics are inventions of the human mind."

Theano wanders closer to look at the computer screen. "That's contrary to what my husband thought. He didn't think we invent math, but rather we *discover* it. Mathematical shapes and numbers are out there in the realm of eternal ideas. They have an independent existence from us. They transcend us and our physical reality."

Julia set fractal.

You nod. "I agree. Let's consider a fractal example. The Julia set for a particular equation has a spiral structure." You point to the screen. "The statement is either true or false. As you can see, it turns out to be true." You turn to Theano. "Was the statement true before the invention and discovery of fractals?"

A stonefly seems to have a particular affinity for your forehead. Theano brushes the fly away. "My husband would say yes," she says. "Fractal Julia and Mandelbrot sets exist whether humans know about them or not."

"I agree. I think mathematics is a process of discovery. Mathematicians are like archeologists."

You reach into an Aubrey hole and bring out a tattered copy of Roger Penrose's book *The Emperor's New Mind*. The gnawed remnant of an acetabulum is stuck to the cover. You toss the acetabulum back into the hole.

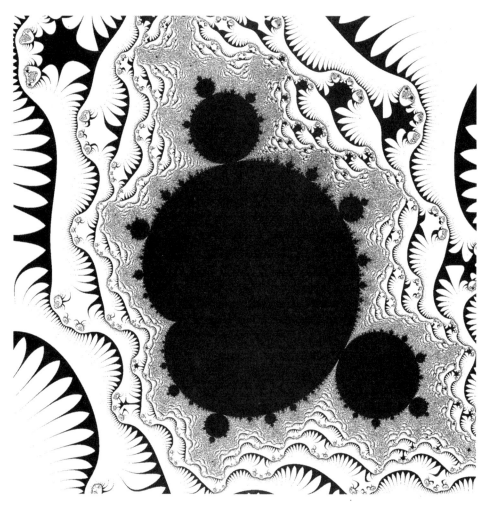

Mandelbrot set fractal.

"Researcher Roger Penrose said that fractals, such as the Julia set or Mandelbrot set, are out there waiting to be found:

> It would seem that the Mandelbrot set is not just part of our minds, but it has a reality of its own ... The computer is being used in essentially the same way that an experimental physicist uses a piece of experimental apparatus to explore the structure of the physical

world. The Mandelbrot set is not an invention of the human mind: it was a discovery. Like Mount Everest, the Mandelbrot set is just there.

You look at Mr. Plex. "I think we are uncovering truths and ideas independent of what computer or mathematical tools we've invented." You pause. "Penrose went a step further about fractals:

> When one sees a mathematical truth, one's consciousness breaks through into this world of ideas.... One may take the view that in such cases the mathematicians have stumbled upon *works of God*.

You say the last three words slowly and distinctly.

Mr. Plex lets out a great gust of wind from his oral cavity, evidently a sound of awe. Theano is quiet and gazes up into the sky.

You follow her gaze. The more recognizable constellations are unmistakable. Orion's square shoulders and feet, the beautiful zigzagging Cassiopeia, and the enigmatic Pleiades all remind you of life back on Earth where you studied the constellations long ago. You even see Aldebaran, the red star in the constellation Taurus. You remember the time you first saw Aldebaran as a boy growing up in Ajaccio, a town in Corsica, the birthplace of Napoleon.

You turn your eyes back to Theano and Mr. Plex. It is December, almost Christmas. Memories flood your mind like a river: memories of the times you've spent with Mr. Plex on your museum ship, memories of your childhood when your mother was still alive.

But mostly you remember the last few Christmases of your life alone. Now you sit by a trilithon, Kylonian wine in hand, wondering where your life is going and why. Well, this Christmas will be utterly different. Theano is with you. You are at Stonehenge. You don't have to know where the road is heading to enjoy the journey.

You are more appreciative of little things. Even the sight of Mr. Plex's glittering eyes and body the color of an insect wing brings a faint smile to your lips.

THE SCIENCE BEHIND THE SCIENCE FICTION

> *I wonder whether fractal images are not touching the very structure of our brains. Is there a clue in the infinitely regressing character of such images that illuminates our perception of art? Could it be that a fractal image is of such extraordinary richness, that it is bound to resonate with our neuronal circuits and stimulate the pleasure I infer we all feel?*
>
> —P. W. Atkins

Fractals

These days computer-generated fractal patterns are everywhere, from squiggly designs on computer art posters to illustrations in the most serious of physics journals. Interest continues to grow among scientists and, rather surprisingly, artists and designers. The word "fractal" was coined in 1975 by mathematician Benoit Mandelbrot to describe an intricate-looking set of curves, many of which were never seen before the advent of computers with their ability to perform massive calculations quickly. Fractals often exhibit self-similarity which means that various copies of an object can be found in the original object at smaller size scales. The detail continues for many magnifications like an endless nesting of Russian dolls within dolls. Some of these shapes exist only in abstract geometric space, but others can be used as models for complex natural objects such as coastlines and blood vessel branching. Interestingly, fractals provide a useful framework for understanding chaotic processes and for performing image compression. The dazzling computer-generated images can be intoxicating, motivating students' interest in math more than any other mathematical discovery in the last century.

Physicists are interested in fractals because they can sometimes describe the chaotic behavior of real-world things such as planetary motion, fluid flow, the diffusion of drugs, the behavior of interindustry relationships, and the vibration of airplane wings. Traditionally when physicists or mathematicians saw complicated results, they often looked for complicated causes. In contrast, many fractal shapes describe the fantastically complicated behavior of the simplest of formulas. The results should be of interest to artists and non-mathematicians, and anyone with imagination and a little computer programming skill. Some readers may wonder why scientists and mathematicians use computer graphics to display mathematical results. Science writer James Gleick says it best:

> Graphic images are the key. It's masochism for a mathematician to do without pictures … [Otherwise] how can they see the relationship between that motion and this. How can they develop intuition?

The Gohonzon of the Soka Gakkai

> *At that time in the Buddha's presence, there was a tower adorned with the 7 treasures, 500 yojanas in height and 250 yojanas in width and depth.… It had 5000 railings, 10,000 rooms, …, 10,000,000,000 jeweled bells.*
>
> —The Lotus Sutra

> *An elephant is contained in a blade of grass.*
> —Ancient Buddhist saying

In Japan, there exists a religious group, the members of which meditate using visual aids resembling fractals. Soka Gakkai is an association of laymen who practice Nichiren-sho-shu Buddhism. The association was founded in Japan in 1937, however the practice of Nichiren-sho-shu Buddhism dates back for centuries to the followers of the Japanese Buddhist saint Nichiren (1222–1282), an ardent nationalist who attacked the established contemporary religious and political institutions of Japan. After exploring different facets of Buddhist Scripture, Nichiren concluded that only one Scripture gave the truth—the *Lotus Sutra* (the Scripture of the Lotus of the Good Law).

One way the Soka Gakkai pay devotion to the Lotus Sutra is through a ritual drawing, or mandala, called the *gohonzon*. The gohonzon contains the names of divinities mentioned in the sutra arranged around the name of the Lotus Sutra. In particular, the gohonzon contains the sacred phrase *Nam-myoho Renge-kyo* (Devotion to the Lotus Sutra) written vertically in the center, around which are arranged names of various deities—perhaps to safeguard the seven-ideograph phrase. Today, millions of followers chant the phrase thousands of times a day.

The gohonzon mandala, an example of which is shown below (left), is placed on an altar in Nichiren temples, as well as in the homes of Nichiren Buddhists. The original *dai gohonzon*—the super-gohonzon for all humanity—resides in a temple in Kyoto. No photographs of the dai gohonzon are allowed, and no pictures have been published. To the Nichiren Buddhists, the gohonzon symbolizes the superiority of the Lotus Sutra over other religions and sects. Nichiren himself drew many gohonzon mandalas which he gave to his followers. Since they were only entitled to the mandalas while they lived, the mandalas were returned to the temples after their deaths.

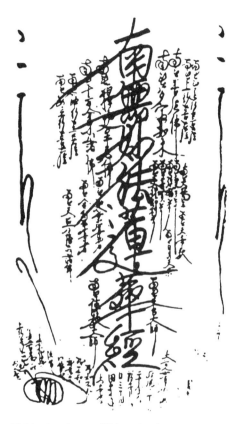

Nichiren's gohonzon (13th century).

Another gohonzon.

The gohonzon reminds me of a fractal, and all the various renditions I have seen have visual details on at least two different size scales. When I talked to Gary Adamson, expert and practitioner of Nichiren-sho-shu Buddhism, he indicated the gohonzon mandala represents the Cosmos, the totality of existence, or in modern mathematical terms, "the class of all sets." Further he believes that the gohonzon's basic theme is: the whole is greater than the sum of its parts. Gary says, "When people chant to the gohonzon, they are not looking at individual parts of the fractal figure, but the complete whole." Gary sent me the gohonzon above (right) from the Nichiren Shoshu sect. The world of Buddhahood resided at the top of the figure, and Hell is below. Buddhism has a classification of 10 dimensions, the topmost dimension being that of enlightened Buddhas.

As with many religions mentioned in this book, Nichiren Buddhism contains various numerical curiosities. For example, one of the main elements of Nichiren's teachings is the theory of *ichinen sanzen* ("1 thought, 3000") which means that the mind, at any moment, contains 3000 different aspects of the universe. To arrive at the figure 3000, Buddhists consider the human mind to be in one of ten states at any particular moment: hell, craving, animalness, anger, tranquility, rapture, intellectual pleasure, truth, mercy and wisdom, and finally Buddhahood. In each state of mind there are the same 10 aspects, bringing the total number of states of mind to 100. There are an additional 10 factors of existence—

Music is frequently a meditative aid in oriental religions. Notice the large drumlike instrument towards the back and center. The drum is bilaterally symmetrical, except for the single arching shape in the middle. The vertical wind instrument at the lower left is called a *sheng* and it contains exactly 13 pipes. I am not aware of other classical instruments with an odd number of elements. (There are 88 piano keys, 4 violin strings, 6 guitar strings, 6 main piccolo holes ...)

appearance, nature, entity, power, action, cause, relationship, effect, reward, and consistency—which bring the 100 states to 1000. Finally, the 1000 is multiplied by 3 to take account of the 3 categories of human life: physical and mental abilities, the relationship between the individual and other members of his community, and the environment in which an individual personality exists.

Makiguchi Tsunesaburo, the founder of Nichiren-sho-shu Buddhism, suggests that the goals of life are beauty, gain, and goodness. Through Nichiren Buddhism, an individual may attain these goals.

Gary Adamson indicates that the ghonozon is unlike other popular visual meditation aids, such as the shri-yantra of India, because the gohonzon is the most free-flowing of all mandalas, having a random component along with a beautiful structure "analogous to many natural phenomena like twigs of a tree." The shri-yantra, an ancient Vedic symbol which represents the seed of the universe—a place that exists beyond space and time—is purely geometric using triangles, squares, circles, and parallelograms. Some medieval and later Hindu temples in India contain shrines with shri-yantra engravings. The shri-yantra is also engraved on foil, placed in a metallic case, and worn as an amulet for health.

The shri-yantra of India, a geometrical diagram used as an aid to meditation. This yantra is composed of nine juxtaposed triangles arranged to produce 43 small triangles. Four of the nine primary triangles point upward, representing male cosmic energy. Five point downward, symbolizing female cosmic power. In southern India, the shri-yantra is an object of worship.

The mouse child uses the image as a yantra to meditate on nothing and infinity.

—Daedalus, 1980

Hermetic Geometry

We who are heirs to three recent centuries of scientific development can hardly imagine a state of mind in which many mathematical objects were regarded as symbols of spiritual truths.

—P. Davis and R. Hersh, *The Mathematical Experience*

Completely involved as he was in Hermetism, Bruno could not conceive of a philosophy of nature, of number, of geometry, of a diagram without infusing into these divine meaning. Bruno based memory on celestial images which are shadows of ideas in the soul of the world, and thus unified the innumerable individuals in the world and all the contents of memory.

—Frances Yates, *Giordano Bruno and the Hermetic Tradition*

Through the ages, various visual patterns such as in the gohonzon have been created to induce spiritual forces to influence material forces. Perhaps the most geometrical of these mystical figures comes from Hermetic geometry where the diagrams represent pure celestial forms. (See *Hermetica* in Postscript 3.) The design on paper was supposed to induce a resonance with its celestial counterpart, and as a result, the figure was thought to have various powers. For example, the symmetrical patterns were used to achieve personal gain, cure diseases, find love, or harm one's enemies. The figure below shows Hermetic designs from 1588 by Giordano Bruno. The Hermetic geometers thought that these designs were keys to the universe.

Giordano Bruno (1548–1600) was an Italian philosopher, mathematician, astronomer, and occultist. Some of his theories anticipated modern science. For example, he rejected the traditional Earth-centered universe and believed in a multiplicity of worlds. He also believed that the Bible should be followed for its moral teaching but not for its astronomical implications.

(*a*) "Figura Mentis."

(*b*) "Figura Intellectus."

(*c*) "Figura Amoris."

(*d*) "Zoemetra."

The Hermetic figures of Giordano Bruno (1588). a) Figura Mentis. b) Figura Intellectus. c) Figura Amoris. d) Zoemetra.

যোগিরাজ শ্রীশ্রীশ্যামাচরণ লাহিড়ী মহাশয়ের প্রদত্ত ষট্‌চক্র চিত্র ও ৪৯ বায়ুর বিবরণ।

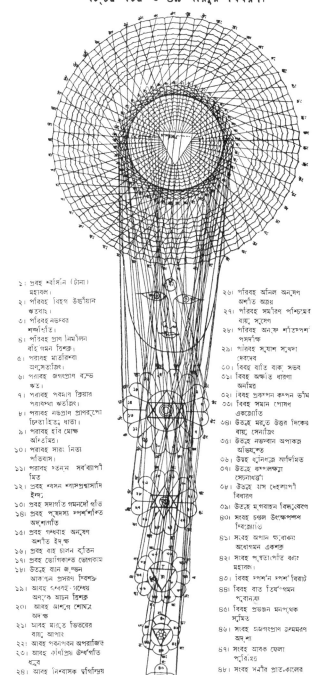

১। প্রবহ শ্বর্সাসন (টানা)
    মহাবল।
২। পরিবহ বিহগ উড্ডীয়ান
    ক্ষতবিঃ।
৩। পরিবহ নভঙ্কর
    শর্বনিঃশ্বত।
৪। পরিবহ প্রাণ নিম্বৌলন
    বাঁহ গমন ত্রিশক্ত।
৫। পরাবহ মাতরিশ্বা
    অঙ্গসেতাঙ্গিতা।
৬। পরাবহ জগৎপ্রাণ বম্ভ্ড
    ঝত্।
৭। পরাবহ পবমান ক্রিয়ার
    পরাবস্থা ক্ষত্তঙ্গ।
৮। পরাবহ নড্ঢপ্রাণ প্রাণপর্পো
    চিতন্নিহিতু ধাতা।
৯। পরাবহ হিব মোক্ষ
    অন্তর্মির।
১০। পরাবহ সারং নিতা
    পাতিবাস।
১১। পরাবহ স্তনন সর্বব্যাপী
    মিত।
১২। প্রবহ শ্বসন শ্বাস্প্রশ্বাসাদি
    ইন্দ্র।
১৩। প্রবহ সদাগাতি গমনেদৌ গাতি
১৪। প্রবহ পৃষদস্যা স্পর্শশারিকা
    অদ্র্শাগাতি।
১৫। প্রবহ গন্ধবাহ অনুষ্ণ
    অশীত ইদ্ক্ষ।
১৬। প্রবহ চালন বর্তান
১৭। প্রবহ ভোগাকান্ত ভোগকাম।
১৮। উত্তবহ বান জ্বলন
    আকঞ্চন প্রসরণ দ্বিশণ্ভু।
১৯। আবহ গন্ধবহ গন্ধেমু
    অন্যেক আচন ত্রিশক্ত।
২০। আবহ আশুগে শোষণে
    অদ্ক্ষ।
২১। আবহ মারুত ভিতরের
    বায়ু আগার।
২২। আবহ পবনপবন অপরাজিত।
২৩। আবহ শোণপ্রিয় ঊর্ধ্বগাতি
    ধ্রুব।
২৪। আবহ নিঃশ্বাসক দুর্গান্ধিমু
    ব্যাপি গাতিব্ঠি।
২৫। আবহ উদান উদ্গীরণ
    সক্ত।

২৬। পরিবহ অনিল অনুষ্ণ
    অশীত অতয়।
২৭। পরিবহ সমীরণ পশ্চিমের
    বায়ু সুষেণ।
২৮। পরিবহ অনুক্ষ শীতুস্পর্শ
    পসদীক্ষ।
২৯। পরিবহ সুযাশ সুখদা
    দেবদেব।
৩০। বিবহ বাতি বাক্ সম্ভব।
৩১। বিবহ অঙ্কিত ধারণা
    অনমিত।
৩২। বিবহ প্রাকল্পন কম্পন ভীম।
৩৩। বিবহ সমান পোষণ
    একজ্যোতি।
৩৪। উত্তবহ মরুত উত্তর দিকের
    বায়ু সেনাত্রাণ।
৩৫। উত্তবহ নভঙ্কবান অপাকজ
    অভিযুক্ত।
৩৬। উত্তবহ ধ্বনিনুজ সর্দিগত।
৩৭। উত্তবহ কম্পলক্ষ্যা
    সেন্দনাধণ্টী।
৩৮। উত্তবহ যাস দেহলাগাগী
    বিধারণ।
৩৯। উত্তবহ মৃগবাহন বিদ্বেষবণ।
৪০। সংবহ চঞ্চল উৎক্ষেপক্ষন
    দ্বিজ্যোতি।
৪১। সংবহ অপান ক্ষুদ্ধাক্ষন
    অধোগমন একশত।
৪২। সংবহ পৃষ্টতাপাতি বলন
    মহাকল।
৪৩। বিবহ স্পর্শন স্পর্শ বিরাট।
৪৪। বিবহ বাত তির্যগ্গমন
    পবনারায়।
৪৫। বিবহ প্রভঞ্জন মনপৃথক
    সমিত।
৪৬। সংবহ হজ্গংপ্রাণ জন্মমরণ
    অদ্শা।
৪৭। সংবহ আবক ফেলা
    পাবিন্ত্র।
৪৮। সংবহ সমীর প্রাতঃকালের
    বায়ু সর্মিত।
৪৯। সংবহ প্রকপ্ন ঈ গতের
    অপ্তর্ যান মহাযান।

The Sahasra chakra, or energy centers, of Tantric literature. The diagram is used to understand how 49 types of breath are associated with the six centers. The thousand petals' center at the head is described as the "door of the Guru" — utter tranquility through inhaling and exhaling. Visually compare the nested geometrical objects with the Hermetic figures of Giordano Bruno. (Drawing by Yogi Lahiri Mahasaya (1828–1895) communicated to Swami Satyeswarananda Giri to Gary Adamason to me.)

দ্রষ্টব্য ঃ (ক) যে সব অক্ষর সহস্রারে রয়েছে সেসব অক্ষর ষট্‌চক্রে রেখাদ্বারা সংযুক্ত রয়েছে।
      (খ) ʇ সহস্রারে তিন 'শ' মূলাধারের 'শ' এর সহিত মিলিত হয়েছে।
      (গ) সহস্রারে লিখিত 'হংসনংযং গুরুর্বৈবঃ' ছবিতে অস্পষ্ট।

Giordano Bruno's patterns are of particular interest because of their geometrical and (often) recursive shapes. Occasionally Bruno adds objects such as serpents and lutes. If you look closely, you can see that the figure ("Figura Amoris") actually has the word MAGIC written in the diagram.

Bruno lived an interesting life. He thought himself a messiah, which was not uncommon for magicians of the Renaissance. The Venetian Inquisition was not pleased with Bruno's ideas and forced him to justify them. He gave the Inquisition a detailed technical account of his philosophy as if he were lecturing great scholars. He told them that he believed the universe to be infinite, because the infinite divine power would not produce a finite world. The Earth was a star, as Pythagoras thought, like the Moon, other planets, and other worlds which were infinite in number.

At the end of the Venetian trial, Bruno fully recanted various heresies of which he was accused and threw himself at the mercy of the judges. By law, his case was sent on to Rome where he was imprisoned as the trial dragged on for years. Finally, in 1599, a judge listed eight heresies taken from Bruno's works which Bruno was required to repudiate. He countered that there was nothing in his work that was heretical and that the church was merely misinterpreting it. He finally declared that he did not even know what he was expected to retract. As a result, Bruno was sentenced as an impenitent heretic. His death sentence was read, and Bruno addressed his judges, "Perhaps your fear in passing judgment on me is greater than mine in receiving it." He was gagged and burned alive in Rome on February 17, 1600.

Unfortunately, today we do not have the report of the Venetian Inquisition, so do not know the 8 heretical propositions which Bruno was required to recant. Scholars believe they have something to do with God's infinity implying an infinite universe, mode of creation of the human soul, motion of the Earth, the soul, multiplicity of worlds, and the desirability of using magic. Other problems arose from the fact that Bruno believed Moses performed his miracles by magic and that Christ was a magician.

Frances Yates, writing in *Giordano Bruno and the Hermetic Tradition*, notes:

> The Renaissance magic was turning towards number as a possible key to operations, and the subsequent history of man's achievements in applied science has shown that number is indeed a master-key, or one of the master-keys, to operations by which the forces of the cosmos are made to work in man's service. However, neither Pythagorean number, organically wedded to symbolism and mysticism, nor Kabalistic conjuring with numbers in relation to the mystical powers of the Hebrew alphabet, will of themselves lead to the mathematics which really work in applied sciences.

Other more recent mystics have also used geometrical diagrams in spiritual realms. For example the figure on p. 159 shows the Sahasra chakra, or energy centers, of Tantric literature. This diagram was drawn in the mid 1800s by Yogi Lahiri Mahasaya (1828–1895).

CHAPTER 14

Behold the Fractal Quipu

Viracocha, Lord of the Universe! Whether male or female, commander of heat and reproduction, being one who even with his spittle, can work sorcery, Where are you? The sun, the moon; the day, the night; summer, winter, not in vain but in orderly succession do they march to their destined place, to their goal.

—Inca prayer to a cosmic god

They have their quipoes, which is a sort of strings of different bigness in which they make knots of several colours, by which they remember.... When they go to confession, these quipoes serve them to remember their sins.
—Churchill's Voyage III, 1704

History has been written with quipo-threads.
—Carlyle, 1857

Theano places some kindling in an Aubrey hole and puts a few cedar sticks and branches on top. "Tetraktys," she whispers as she throws a match into the pit. In a few seconds, the kindling is crackling.

Theano stares into the fire as it grows brighter, tongues of fractal flames licking the dried bark of the branches.

The fire is blazing now, the delicious aroma comforting you as the skeletal branches are engulfed in bright amber flames. But you feel as if you are in a trance. Even your own breathing seems too slow and strange. And you aren't sure now that your voice is audible.

Slowly Mr. Plex sits down on the warm sod beside a large Stonehenge rock. He peers into the shadows beneath the arch of the trilithon.

You hear Theano sighing again, but it is soft and quick, like the wind. She rests against the stone, basking in the heat of the fire, her eyes wide as she gazes into the play of lights on the rocks. It seems she has been here forever.

You break out of your trance and toss a colored, knotted cord to Mr. Plex.

He catches it and studies it with his multiple limbs like a spider spinning a web. "What is it?"

"It's a quipu. From the ancient Incas." You pronounce it "kee-poo."

Mr. Plex hands it to Theano who stares at the complex arrangement of knots.

"Incas?" she asks.

"Five million people who lived from A.D. 1400 to A.D. 1560 in what is now Peru, and also in parts of modern Ecuador, Bolivia, Chile, and Argentina."

Seeing the confusion in Theano's face, you sketch a map of the world in the sand and point to South America.

"They had a developed civilization—a common state religion, a common language. Although they didn't have writing, they kept extensive records encoded by a logical-numerical system on the quipus."

Theano begins to play with the intricate arrangement of cords. "Why'd they need to keep complicated records?"

You toss a branch into the fire. "The Incan governmental bureaucracy kept careful records on the goods they redistributed from their store-houses to their army. They encoded and de-coded a tremendous amount of information using the quipus."

Theano begins to count the knots in the quipus.

Your tone turns sad. "Unfortunately, when the Spanish came to South America, they saw the strange quipus and thought they were the works of the devil. The Spanish destroyed thou-sands of them in the name of God, and today only about 500 quipus remain."

Theano holds up the quipu. "Where'd you get this one?" The quipu's multiple cords trail beneath her slender hand like a living octopus.

"I had an excavation mole locate one un-derground. All of the quipus we have today were recovered from graves, probably buried with those who made them."

A shudder runs up Mr. Plex's abdomen. "I hate those excavation moles. Once one crawled down my alimentary canal. Had a hell of a time trying to get it out."

Inca holding a simple knotted quipu. At the lower left is a binary colored counting device with a "memory" of 20 compartments.

Theano turns to you. "How does it work?"

"A quipu has a main cord from which the other cords are suspended." You point to the large primary cord. "Most of the attached cords fall in one direction and they're called *pendant cords*. A few lie in the opposite direction and are called *top cords*."

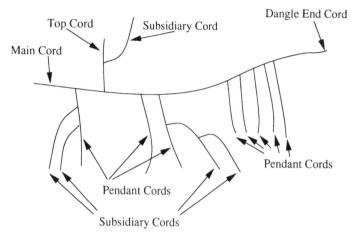

Theano pulls on a cord suspended from the pendant cords. "What do you call this one?"

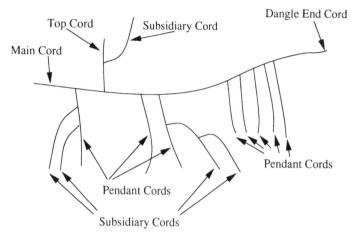

"*Subsidiary cord.* There can be subsidiaries on the subsidiaries forming a fractal branching pattern."

You sketch in the sand.

You stand up and stretch your legs. "My excavation mole once unearthed a quipu with 18 subsidiaries on one level. The quipu had 10 levels of subsidiaries! The largest quipu found contained 1000 cords. The smallest 3 cords."

Theano tugs at a knot. "What do the knots mean?"

"Knot types and positions, cord directions, cord levels, and color and spacing are all interpreted by the quipu makers to represent numbers." You pause. "The knots were probably used to record human and material resources and calendric information. Some archeologists think that they contained more information: construction plans, dance patterns, and even aspects of Inca history."

Theano nods.

"One sinister application of the quipu was as a death simulator."

Theano wrinkles her brow. "What the Hades is that?"

"Yearly quotas of adults and children were ritually slaughtered. The whole thing was planned using a quipu—a cord model of the entire empire. The cords represented roads, and the knots were sacrificial victims."

"Sick," Theano whispers.

The fire is dying. Mr. Plex pokes at some of the burning branches creating a fountain of sparks. "Sir, did the Incas believe in God?"

You nod. "They believed that the creator resided in Tiahunaco where he turned men and women into stone for not obeying his commands. The creator was *Ticci Virachocha,* meaning the unknowable God. It was the Inca name for the ancient Peruvian sun god."

Theano places the fractal weaving upon her neck. "We would make a killing selling these to the Pythagoreans."

You smile as the number necklace dances upon her neck like a heavenly web.

"It's getting mighty cold," Theano says. "Look." She holds out her arm. Flakes of snow land on her Greek robe.

"We had better get out of here," you say. "This isn't a good place to be in a snowstorm." You gaze at the mutant Dionysus. "I think it's time to leave Stonehenge."

THE SCIENCE BEHIND THE SCIENCE FICTION

The Incas really used quipus, ancient memory banks for storing numbers, and the Spaniards really did destroy any quipus they found fearing they were the works of the devil.

THE LOOM
OF GOD

Sixteenth century drawing of a simple quipu. Does the figure hint
at some astronomical purpose for quipus?

A quipu in the collection of the Museo National de Anthropologia y Arquelogia, Lima, Peru. Photo by
Marcia and Robert Ascher.

Also, the quipu human sacrifice simulator is well documented. The top figure on p. 164 is a 16th century drawing of a simple quipu, and the bottom figure is a photograph of an actual quipu.

Sometimes knots had different meanings, depending on which string they were located. For example, a knot on one string might stand for 10, a knot on another might stand for one. Did the quipus code for a certain astronomical period? Cosmic, magical numbers? The population of a village? All these applications are possible.

The Incas also used a kind of primitive computer. It resembled a box with 20 compartments placed in four rows of five (see p. 162). Black and white stones were placed in the compartments, and a compartment was considered filled when it contained 5 stones. In 1590, Padre Jose de Acosta watched the Incas manipulate this abacus-like device, and he drew a sketch. Unfortunately, the Padre could not decipher how the device worked. Today, we still are not sure of its use.

The Appendix contains a C program listing for producing points on a simple quipu. The points can be used to draw a quipu such as the one shown below.

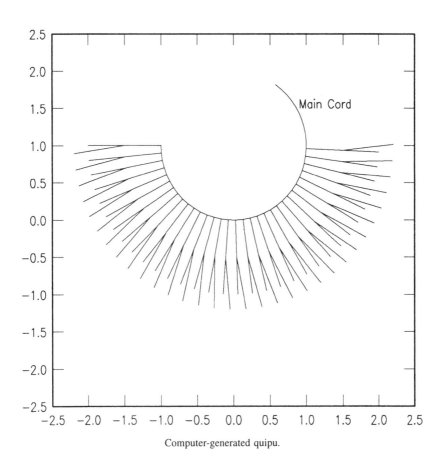

Computer-generated quipu.

數 數 數 數

In this chapter, we've touched again upon *ethnomathematics*, the study of mathematics which takes into consideration the culture in which mathematics arises. Mathematics is associated with the study of "universals." However, what we think of as universal is often colored by our *cultural* and *historical* perspectives. Various researchers believe that mathematics can be studied as a cultural development of structures and systems of ideas involving number, pattern, logic, and spatial configurations. Even the most basic of human activities often rely upon mathematics; these activities include: architecture, weaving, sewing, agriculture (requiring calendar systems and layouts for fields), kinship relations, ornamentation (such as tilings in the Islamic world and Native American beadwork), and spiritual and religious practices which are often associated with patterns in nature or ordered systems of abstract ideas.

The quipu was used around 1000 years ago. This means that the quipu is one of the most ancient methods of useful codification in the New World. The Incas had a highly ordered civilization as exemplified by their quantification of the planting of crops, behavior of the armies, output of gold mines, census results, composition of work forces, amount and kind of tributes, the contents of storehouses—and all these items were all recorded on the quipus.

When power was transferred from one Inca emperor to the next, information stored on quipus was used to recount the accomplishments of the leaders. Interestingly, today there are computer systems whose file managers are called quipus, no doubt in honor of this very useful and ancient device.

CHAPTER 15

The Eye of God

The power of the Golden Section to create harmony arises from its unique capacity to unite the different parts of a whole so that each preserves its own identity, and yet blends into the greater pattern of a single whole. The Golden Section's ratio is an irrational, infinite number which can only be approximated. Yet such approximations are possible even within the limits of small whole numbers. This recognition filled the ancient Pythagoreans with awe: they sensed in it the secret power of cosmic order. It also led to the endeavors to realize the harmonies of such proportions in the patterns of daily life, thereby elevating life to an art.

—Gyorgy Doczi, *The Power of Limits*

Even hard-nosed atheists frequently have what has been called a sense of reverence for nature, a fascination and respect for its depth and beauty and subtlety that is akin to religious awe.

—Paul Davies, *The Mind of God*

"Theano, what are you drawing?"

"The magic symbol of the Pythagoreans. We usually try to keep it secret, but I've nothing to hide from you and Mr. Plex."

Theano has sketched a five-pointed star, or pentagram, on the sand.

The three of you are standing beneath a trilithon, sheltered from most of the snow. All your belongings are packed, and you are ready to leave Stonehenge. Fat flakes of snow descend from the sky. The flowers around Stonehenge droop, soggy and heavy.

Mr. Plex stares at the pentagram. "Sir, so many religions and cults have used this symbol."

You nod. "It's the pattern Goethe's Faust used for trapping Mephistopheles. Turn it upside down, and it's a satanic symbol."

You study the figure for a few seconds and turn to Theano. "Did you know that every line segment in the pentagram is in divine proportion to every segment of the next smaller

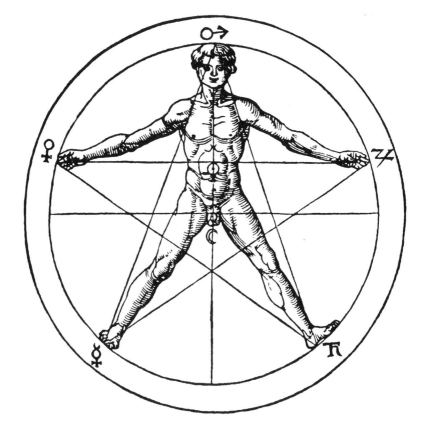

Symbolic pentagram representation of man as microcosmos (from Agrippa von Nettesheim, 1535).

length. Also, the diagonal of the regular pentagon inside the pentagram is in golden proportion to the pentagon's side."

Theano looks up at you. "Divine ratio?"

"It's a ratio which appears with amazing frequency in mathematics and nature. It's sometimes called the divine proportion or golden ratio."

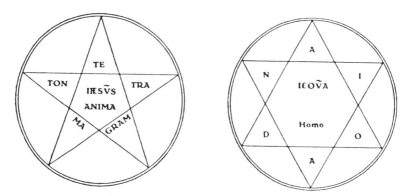

The Pentaculum (left) and Sigillum (right). Paracelsus writes in *De Occulta Philosophia*: "With these two signs the Israelites and the necromantic Jews have done much and brought about much. They are still kept highly secret by a number of them. For these two have such strong power that everything that can done by characters and words is possible for these two."

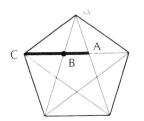 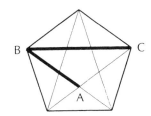 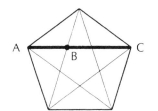

You draw a line on the ground.

"I can explain the proportion most easily by dividing a line into two segments so that the ratio of the whole segment to the longer part is the same as the ratio of the longer part to the shorter part: AB/AC = AC/CB = 1.61803...."

You draw a rectangle on the ground.

"If the lengths of the sides of a rectangle are in the golden ratio, then the rectangle is a golden rectangle:

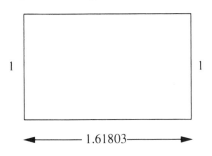

Now watch this. Let's cut off a square inside the parent rectangle." You sketch a square and fill it with a brick pattern.

"This cutting forms another smaller golden rectangle within the parent. We can then cut a square from the baby golden rect-

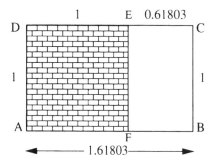

angle and again we are left with another smaller golden rectangle. We could continue this process indefinitely, producing smaller and smaller golden rectangles."

"Sir, where does it all end?"

"If you draw a diagonal from the top left of the original rectangle to the bottom right, and then from the bottom left of the baby golden rectangle to the top right, the intersection point shows the point to which all the baby golden rectangles converge."

"Mon Dieu!"

"And the diagonals' lengths are in golden ratio to one another!" You pause. "In honor of the various 'divine' properties attributed to the golden ratio over the centuries, my colleagues call the point to which all the golden rectangles converge 'The Eye of God.' We can keep magnifying the figure but can never get to the Eye using finite magnifications."

You look from Mr. Plex to Theano. "The golden rectangle is the *only* rectangle from which a square can be cut so that the remaining rectangle will always be similar to the original rectangle."

You begin to sketch a curve on the

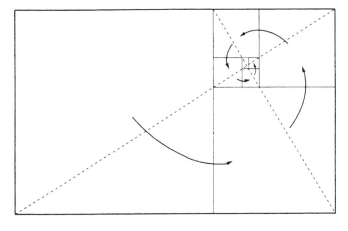

nested golden rectangles. "If you connect the vertices, you form a logarithmic spiral that 'strangles' the Eye of God. Logarithmic spirals are everywhere—seashells, horns, the cochlea of the ear—anywhere nature needs to fill space economically and regularly. A spiral is strong and uses a minimum of materials. While expanding, it alters its size but never its shape."

Theano's brow is knitted together in concentration. "That sounds like a self-similar fractal."

"Right, I told you about self-similarity. The logarithmic spiral is self-similar because you can magnify and rotate any portion to produce any part of the spiral. However, this is a *smooth* self-similar object in stark contrast to the fractals we usually associate with self-similarity such as jagged coast-lines."

Mr. Plex is doodling in the snow with his right forelimb. "Sir, did any religions make use of the golden rectangle?"

"In 1509, Fra Luca Pacioli published *De divina Proportione* illustrated by Leonardo da Vinci who was the first to refer to the proportion as the *sectio aurea*, the Golden section. Pacioli presented 13 of the golden ratio's remarkable properties. He stopped at 13 because 13 was the number of people at the table of the Last Supper. Pacioli said, "for the sake of salvation, the list must end here." You pause. "Kepler went wild over the Divine Proportion, saying

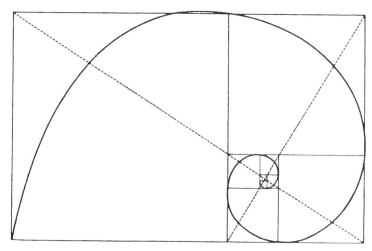

A logarithmic spiral can be generated by nesting squares and golden rectangles. The spiral swirls to an intersection point of the two diagonals in the figure. The diagonals themselves are in the golden ratio to one another. In honor of the various divine properties attributed to the golden ratio, we may wish to call this intersection point, "The Eye of God." (After Gardner, 1992.)

'Geometry has two great treasures: one is the Theorem of Pythagoras, the other the division of a line into [the golden] ratio. The first we may compare to a measure of gold, the second we may name a precious jewel.' "

You take a deep breath and brush away some snow that has accumulated on the mutant Dionysus. "But today we're more interested in the golden ratio's *mathematical* properties. For example, the golden ratio is also related to the Fibonacci sequence in an amazing way."

Theano's eyes begin to watch you with renewed interest. "Fibo-what?" she says.

"Fibonacci. Like the golden ratio, Fibonacci numbers are seen in nature—for example, in arrangement of plants' leaves, buds and seeds. This sequence of numbers is named after the wealthy Italian merchant Leonardo Fibonacci of Pisa." You pause. "Theano, reach into the Aubrey pit."

With a cold and trembling hand, she pushes a damp strand of hair away from her face, and plunges her hand into the hole. She pulls out a sunflower.

You point to the flower in her hand. "Sunflower heads, like other flowers, contain two families of interlaced spirals—one winding clockwise, the other counter clockwise. Why don't you count the number in a spiral?"

Theano points to the seeds as she counts. "34 in one direction, 55 in the other."

"Aha, Fibonacci numbers! Do you see? The numbers arising in plants are special. Lilies have 3 petals. Buttercups have 5, marigolds 13, asters 21, and most daisies have 34, 55, or 89."

Mr. Plex scratches his abdomen. "I've seen these numbers before."

"The Fibonacci series is: 3, 5, 8, 13, 21, 34, 55, 89...."

"Sir, each number is the sum of the two that precede it!"

"Weird," Theano says.

"Right. And the numbers arise when you consider the breeding of rabbits." You pause. "You go to a neighborhood pet store and buy a pair of small rabbits and breed them. The pair produces one pair of young after one year, and a second pair after the second year. Then they stop breeding. Each new pair also produces two more pairs in the same way, and then stops breeding. How many new pairs of rabbits would you have each year? To answer this question, write down the number of pairs in each generation. First write the number 1 for the single pair you bought from the pet shop. Next write the number 1 for the pair they produced after a year. The next year both pairs have young, so the next number is 2. Continuing this process, we have the sequence of numbers: 1, 1, 2, 3, 5, 8, 13, 21, 34, 55, 89, 144, 233, 377,... This sequence of numbers, called the *Fibonacci sequence*, plays important roles in mathematics and nature. These numbers are such that, after the first two, every number in the sequence equals the sum of the two previous numbers $F_n = F_{n-1} + F_{n-2}$."

"But sir, how does all this relate to the golden ratio?"

You smile. "The neat thing is that the ratio of two successive Fibonacci numbers approximates the golden ratio. You could even construct approximately golden rectangles by using two adjacent Fibonacci numbers, like 55 and 34, for the lengths of two sides." You pause. "The golden ratio appears so often in mathematics that some have used it as evidence for the existence of a geometer God ..."

THE SCIENCE BEHIND THE SCIENCE FICTION

The Cult of the Golden Ratio

Although the golden rectangle is said to be the most visually pleasing of all rectangles, being neither too squat nor too thin, I find little evidence for this. In addition, many artistic works are claimed to contain examples of the golden ratio, for example: the Greek Parthenon, Leonardo da Vinci's *Mona Lisa*, Salvador Dali's *The Sacrament of the Last Supper*, and much of M.C. Escher's work. Again, the evidence for this appears to be thin. Martin Gardner refers to this collection of beliefs, and promoters of such beliefs, as "The Cult of the Golden Ratio." George Markowsky from the University of Maine notes that the golden rectangle does not score any better in terms of aesthetics than rectangles similar to it, such as a 3×5 file card.

However, no one doubts that the golden ratio has some rather amazing mathematical properties. It's symbolized by ϕ, the Greek letter phi. Since $\phi = (1 + \sqrt{5})/2$, one can show that

$$\phi - 1 = \frac{1}{\phi}; \quad \phi\phi' = -1; \quad \phi + \phi' = 1; \quad \phi^n + \phi^{n+1} = \phi^{n+2}$$

where $\phi' = (1 - \sqrt{5})/2$. Both ϕ and ϕ' are the roots of $x^2 - x - 1 = 0$. In addition, we have the remarkable formulas involving the single digit, 1.

$$\phi = \sqrt{1 + \sqrt{1 + \sqrt{1 + \sqrt{1 + \cdots}}}}$$

$$\phi = 1 + \cfrac{1}{1 + \cfrac{1}{1 + \cfrac{1}{\cdots}}}$$

This second formula is an elementary example of continued fractions which have numerous uses for mathematicians and physicists. A general expression uses the letter b to denote the number in each denominator in the previous formula.

$$\phi = b_0 + \cfrac{1}{b_1 + \cfrac{1}{b_2 + \cfrac{1}{\cdots}}}$$

Manfred Schroeder (1986) remarks, "Continued fractions are one of the most delightful and useful subjects of arithmetic, yet they have been continually neglected by our educational factions." These typographical nightmares can be more compactly written as:

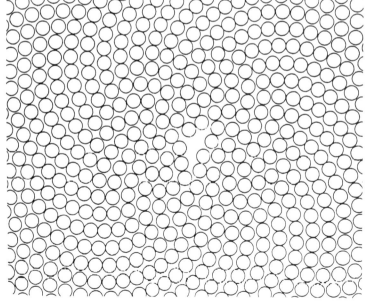

Computer-generated Fibonacci spirals of circles.

$[b_0, b_1, b_2, \ldots]$. And the golden ratio can be represented as $[1,1,1,1, \ldots]$. We can also write this more compactly as: $[1,\overline{1}]$, where the bar indicates a repetition of the number 1. It is mind-boggling that the irrational number e ($e = 2.718281828 \ldots$), unlike π, can be represented as a continued fraction with unusual regularity $[2,1,2,1,$ $1,4,1,1,6,1,1,8,1, \ldots]$, however this converges very slowly to the true value of e because of the many 1s. In comparison, the golden ratio, which contains infinitely many 1s, is the most slowly converging of all continued fractions. Schroeder notes, "It is therefore said, somewhat irrationally, that the golden section is the most irrational number." The approximation of the continued fraction to the golden ratio is worse than for any other number. Therefore, chaos researchers often pick the golden ratio as a parameter to make the behavior of simulations as aperiodic as possible.

There are so many deep mathematical expressions, characteristics, and "coincidences" involving the golden ratio and its close associates, the Fibonacci numbers, that entire books and journals are devoted to them. It's no wonder that the cult of the golden ratio has grown in recent decades. (The Appendix contains a BASIC program listing for computing Fibonacci numbers.)

Make Your Own Fibonacci Spiral

Fibonacci spirals, such as those shown on p. 172 are involved in the various plant structures, for example, in the arrangement of seeds in a sunflower head. We can approximate a Fibonacci spiral pattern by drawing circles which have polar coordinates with radii $r_i = k\sqrt{i}$ and angles $\theta_i = 2i\pi/\phi$ where ϕ is the golden ratio $(1 + \sqrt{5})/2$. Radii increasing as the square root of the integers create a mean density of packing that is constant.[1]

The golden ratio pops up in the most unlikely of places. Shown at right is an interesting example where the radius of the large circle divided by the diameter of one of the small circles is the golden ratio.

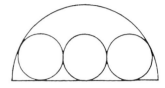

The golden ratio pops up in the most unlikely of places. Here the radius of the large circle divided by the diameter of one of the small circles is the golden ratio.

Voting in Utopia

Imagine a race of perfect beings, maybe angels, who vote on various matters of spiritual importance. What proportion of votes in favor of an issue should be considered as a mandate from the angels? One possible answer is the golden ratio. In this way, the ratio of all the votes to the "yes" votes is the same as the ratio of the "yes" votes to the "no" votes. In another words, if 1.618 times as many angels vote yes rather than no, then a resolution is passed.

Joe McCauley from Colorado has had similar thoughts:

> On U.S. election nights when TV networks tally the voting results, newscasters use terms such as "landslide" or "mandate" when a candidate wins by a sizeable margin. I find myself wondering what is the minimum required margin of victory to qualify as a "mandate." In two-candidate elections two-thirds of the vote seems like more than enough, while three-fifths doesn't seem like quite enough. In my opinion, the winner's portion of the total vote should be greater than the loser's vote as a portion of the winner's vote. This works out to be about 61.8%, which corresponds to the reciprocal of the golden ratio and one less than the golden ratio.

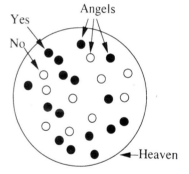

A golden majority of "yes" angels (black dots). The ratio of all the votes to the "yes" votes is the same as the ratio of the "yes" votes to the "no" votes. In another words, if 1.618 times as many angels vote yes rather than no, then a resolution is passed.

Dr. George Markowsky from the University of Maine writes to me:

> In some correspondence, mathematician Donald Knuth notes that the Bible uses a phrase worded something like "as my father is to me, I am to you" which suggests the basic equation of the golden ratio (because if I divide a line into two segments so that the ratio of the whole segment to the longer part is the same as the ratio of the longer part to the shorter part, I get the golden ratio.) This is not a proof that God looks favorably upon the golden ratio but should be of interest to golden ratio aficionados.

In 1994, the *Journal of Recreational Mathematics* reported the following enigmatic equation uniting the Number of the Beast 666 (Chapter 5) and the golden ratio ϕ

$$\phi = - \{\sin 666° + \cos [(6)(6)(6)°]\}$$

(See "The Sign of the Devil," page 203, volume 26, issue 3.)

Spirals in Hyperspace

Theano asked whether a logarithmic spiral is a fractal. Perhaps the best and most intriguing answer comes from Dave Uherka (University of North Dakota Mathematics Department). His explanation will appeal to those of you who have mathematical interest, particularly in the area of fractal dimensions, which describe the way a curve fills space:

> A "logarithmic spiral" can be created by $r = \exp(-\theta)$ in polar coordinates. The spiral is self-similar at the origin and has a Hausdorff and box dimension of 1. However, other spirals can have any box dimension desired from 1 to 2. For example, the spiral about the unit circle, $r = 1 + 1/\theta^p = 1 + \exp(-p \ln(\theta))$, where $p > 0$, has box dimension $1 + 1/(p + 1)$, but it has Hausdorff dimension 1. The spiral $r = 1 + 1/\ln(\theta) = 1 + \exp(-\ln \ln(\theta))$ has box dimension 2 and Hausdorff dimension 1. None of these other spirals are exactly self-similar at the origin as is the case with the logarithmic spiral. Instead, they spiral around the circle $r = 1$. If we omit the initial term "1" from the last two spirals, then they spiral about the origin, but I believe that they then have box dimension 1 rather than $1 + 1/(p + 1)$ and 2 respectively.
>
> Since there is no precise definition of fractal—some use the notion of self-similarity while others use the notion of fractional dimension—there will always be arguments as to whether an object is a fractal. Even if one decides to define the term fractal using dimension, some people use Hausdorff dimension and some use box dimension, and these two values are not always equal.
>
> Perhaps this is why Mandelbrot finally gave up on trying to define a fractal, and Barnsley decided to call nearly everything a fractal; that is, his space of fractals is the set of all compact subsets of a complete metric space. This includes all of our classical geometric objects such as line segments, triangles, and circles.

(David warns that the dimension information is tentative and a subject for further research.)

When I asked Dave Uherka to describe the visual appearance of his unusual spirals with various fractional dimensions, he responded:

> The spirals about the origin defined by the formulas $r_1 = \exp(-\theta)$, $r_2 = 1/(\theta)^p$ (as p decreases toward 0), and $r_3 = 1/(\ln \theta)$, all wind around the origin at different rates (r_1 fast, r_2 medium, r_3 slow). For each spiral I assume θ increases from 0 (or 2π in the last case) and approaches $+\infty$. Proceeding from r_1 to r_3, the graphs become denser and denser in the vicinity of the origin. In the last case, the graph near the origin is completely filled in like a black ball drawn on a white screen. It is easy to plot the spirals on a computer, except that in the latter cases (with small p or $r_3 = 1/(\ln \theta)$), the plot may develop so slowly that it may take months to finish the graph. Therefore, one has to finish it by filling a circle near the origin after the spirals are so close together that adjacent pixels are being filled several times. The first spiral, $r_1 = f(\theta) = \exp(-\theta)$, has the property $f(\theta + 2\pi) = f(\theta) \times \exp(-2\pi)$, which means that each loop around the origin is a scaled down version of the previous loop. This gives the spiral self similarity around the origin.

Star-shaped symbols, such as the ones in this chapter, have played significant roles in religion. My favorite pentagram in religion comes from the Bab, an important prophet of the Baha'i religion which had its roots in 18th century Iran. Shown below is a star *haykal* (body or form) in the Bab's handwriting. A common form of Babi scripture is that represented by amulets or talismanic devices, usually drawn in the form of stars. The Bab gave instructions for his followers to wear about their necks a haykal in the Bab's handwriting. The Bab also spoke of haykals consisting of 2001 names of God to be worn from birth as an amulet and never left off. Other haykals consisted of phrases from the Khoran.

毅 毅 毅 毅

Despite the huge diversity of nature, mathematics and science attempt to reduce this complexity to a few general principles. In this chapter, I've given hints that one enigmatic number, the golden ratio ϕ, appears and reappears throughout nature and mathematics. Other researchers, like Professor Jay Kappraff of the New Jersey Institute of Technology, have even discussed the possible use of the golden ratio in understanding crystals and the music of Bela Bartok (1881–1945), a famous Hungarian musician. The frequent appearance of the golden ratio in many diverse branches of geometry leads many to be amazed and delighted that the universe appears united in a wondrous way. As general mathematical theories are

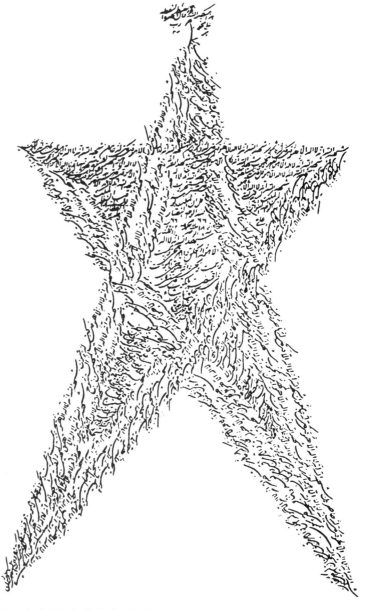

A star haykal in the Bab's handwriting.

derived to encompass these disparate occurrences of the golden ratio, some of this sense of wonder will melt away, to be replaced by intellectual pleasure and hunger for new research and understanding.

CHAPTER 16

Number Caves

If you were a research biologist from the Stone Age, and you had a perfect map of DNA, could you have used it to predict the rise of civilization? Would you have foreseen Mozart, Einstein, the Parthenon, the New Testament?
—Deepak Chopra, *Ageless Body, Timeless Mind*

Was there meaning in the geometry, magic beyond magic? Was there a cave rabbi who could read the kabala?"
—Gerald Hawkins, *Beyond Stonehenge*, 1973

La Pileta was more numerical than artistic, more mystical than real. If rows of dots, strings of vertical dashes, and pairs of strokes were numerical, then La Pileta was a number cave.
—Gerald Hawkins, *Beyond Stonehenge*, 1973

You shine the light into the crevice. "Follow me. We're going to look at something made around 20,000 B.C." The surface of the cave walls are tan, glittering with gypsum crystals. The air smells clean and wet, like hair after it is freshly shampooed.

Huddled together like little hobbits, the smaller stalagmites of calcite cluster near a clear pool. The larger ones looked like rib bones of some giant prehistoric creature.

"Sh–Shall we go in?" Mr. Plex asks.

The wind sings through the stalactites like a large bird.

You put one foot into the crevice. "Sure, I have something wonderful to show you." The cave throws your voice back at you, hollow and spooky.

You shine the flashlight all around. "Incredible," you say. Glittering blue gypsum chandeliers, at least 25 feet in length, are suspended above your heads. Walls encrusted with fragile violet aragonite "bushes" line your path. With just a few steps, you have entered another world.

Theano walks over to a shimmering lake. A gentle breeze paints swirls and eddies on its surface.

You follow, careful not to lose your footing on the slippery cave floor.

The lake has a dank odor and the smell of life. Could any fish live in such a place? You walk closer to the edge of the lake. "Brrr," you say with a shiver. The air is so cold that your breath sometimes steams. "Must be 100 percent humidity."

Your gaze drifts to the pockets of crystals surrounding you. There is a flowing harmony of fractal formations and crystalline outcropping of rock coated with strips of velvet purple. A cool peace floods you, and for a second your concern about the transfinites and your preoccupation with the End of the World ebb away.

Theano places her finger in the lake. It is perfectly black down below.

You look off over the water. It is clear now, in the light of the flashlight, and filled with quartz nodules. You blow on the water, and countless ripples appear on its surface.

"Follow me."

Mr. Plex's forelimbs are quivering. "Sir, it's a bit dark in here."

You lead Mr. Plex and Theano deeper into the cave, several hundred yards into limestone rock, and then a hundred feet down.

Theano turns back to you. "Where are we anyway?"

"The cave of La Pileta, 40 kilometers from the Spanish Mediterranean."

Upon scaling a 5-foot-high dirt wall in the middle of the passageway, you discover piles of human skulls scattered across the ground.

"Look," Theano says as she points to limb bones stacked in piles.

You stoop down. "Interesting. It suggests that the cave was a *secondary* burial sight. The bodies were probably first buried elsewhere until the flesh decayed, then taken to the cave for permanent burial."

"Creepy," Theano whispers.

"Sir, I'd love to gather some bone and charcoal samples for radiocarbon dating back at our ship." Mr. Plex brings out a backpack and stuffs a leg bone into it.

You nod. "Maybe we can do some DNA studies on the bones to figure out if the burial contained dead from the same households."

You shine the flashlight on the wall. "Ah, now here's why I brought you!" You point to parallel marks, groups of five, six, or higher numbers. Clusters of lines are connected across the top with another line, like a comb, or crossed through in a way reminding you of the modern way of checking things in groups of five.

Theano goes closer to the wall. "Were they counting something?"

"I'm not sure. This certainly contradicts old-fashioned notions that cavemen of this period made guttural noises, and were only concerned with feeding and breeding. If the people who drew these designs mastered numbers, cavemen had intellects beyond the minimal demands of hunting." You pause. "Higher thinking."

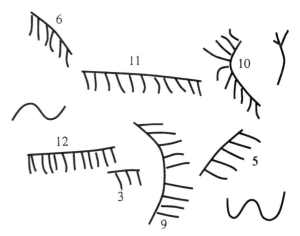

Designs on the wall of the Number Cave. Some researchers believe the markings represent numbers. If you were to explore the cave and consider the teeth of the "combs" as units, you could read all numbers up to 14. In one area of the cave, the numbers 9, 10, 11, 12 appear close together. Could it be that the artist was counting something, recording data, or experimenting with mathematics?

Mr. Plex taps his forelimb on the wall. "Still they could just be abstract designs."

"Correct, but if we still regarded Mayan friezes and decorated pyramids as merely art, we'd be wrong. Luckily, mathematically minded scholars studied them and discovered their numerical significance."

Theano points to an abstract design. "What's this?"

"No one knows. A house with arched doorways? Flat feet? An astronomical symbol?"

Theano shivers. "Let's go to the mouth of the cave and make a fire."

You nod. "Good idea."

After a few minutes Theano has gathered dead spruce branches festooned with lichen. You toss a match, and soon a warm fire is burning.

You close your eyes for a second. Even the faint dripping sound is music. You have longed for this fragrant and embraceable coolness with your soul. Is there anyplace else in Spain, the world, where the moist air feels like a living presence, where the cold air brushes against you like cats, where the stalactites and slippery cave walls are shimmering and alive?

Can Theano feel what you are feeling? Can she appreciate the majesty and mystery of this Stygian chamber? You have the urge to squeeze her hand, hold her close. You've always fallen for the helpless, the ones who were lost or in trouble. And you always regretted getting involved. Why do you always insist on rescuing a damsel in distress? This doesn't seem like the time for romance. Theano is too vulnerable, from a different time, worried about the world. What kind of man could think of her in romantic terms at a time like this? You feel a twinge of shame. But at least you are together....

Her eyes are unfocused as she trails her fingers along a stalagmite.

"What are you thinking?" you ask.

"The things you've shown me. The fractals, the quipu, this number cave. Stonehenge, the golden ratio, the Urantia, perfect numbers. Humans have always intertwined mathematics with the divine."

You nod. "Mathematics is the loom upon which God weaves the fabric of the universe."

THE SCIENCE BEHIND THE SCIENCE FICTION

Number Caves

The cave of La Pileta is 40 kilometers from the Spanish Mediterranean and discussed by Hawkins (see References). The comblike markings you showed to Theano and Mr. Plex really do cover the wall. Some archeologists believe that the markings represent numbers, but we cannot be certain.[1]

The arched doorway diagrams that Theano discovered exist, but their meaning is unclear. What do you think they represent?

The Loom of Computers

Cave wall numbers were some of the initial steps toward primitive computing machines. One of the first true calculating machines to help expand the minds of humans was the abacus. The abacus is a manually operated storage device which aids a *human* calculator. It consists of beads and rods, and originated in the Orient more than 5000 years ago. Archeologists have since found geared calculators, dating back to 80 B.C., in the sea off northwestern Crete. Since then, other primitive calculating machines have evolved, with a variety of esoteric sounding names, including: Napier's bones (consisting of sticks of bones or ivory), Pascal's arithmetic machine (utilizing a mechanical gear system), Leibniz' Stepped Reckoner, and Babbage's analytical engine (which used punched cards).

Continuing with more history: the Atanasoff-Berry computer, made in 1939, and the 1500 vacuum tube, Colossus, were the first programmable electronic machines. The Colossus first ran in 1943 in order to break a German coding machine named Enigma. The first computer able to store programs was the Manchester University Mark I. It ran its first program in 1948. Later, the transistor and the integrated circuit enabled micro-miniaturization and led to the modern computer which today is of central importance to many researchers exploring the mathematical loom of the universe.

In the mid-1990s, one of the world's most powerful and fastest computers was the special purpose GRAPE-4 machine from the University of Tokyo. It achieved a peak speed for a computer performing a scientific calculation of 1.08 Tflops. (Tflops stands for trillion floating point operations per second.) With this computer, scientists performed simulations of the interactions among astronomical objects such as stars and galaxies. This type of simulation, referred to as an N-body problem because the behavior of each of the N test objects is affected by all the other objects, is particularly computation intensive. GRAPE-4 reached its record speeds using 1692 processor chips, each performing at rates of 640 Mflops. Like a web spun by a mathematically inclined spider, each processor had intricate connections with one another. The Tokyo researchers hoped to achieve petaflops (10^{15} operations per second) by the turn of the century with a suite of 20,000 processors each operating at 50 Gflops. GRAPE-4 is certainly much more expensive than the abacus or Napier's bones, but also much faster!

CHAPTER **17**

Numerical Gargoyles

Friends and dragons on the gargoyled eaves,
Watch the dead Christ between the living thieves.
  —Longfellow, *Divina Comm.*, Sonnet ii, 1864

Out of the Mouthes of certain beastes or gargels did runne red, white, and claret wine."

—Chronicles, 1548

A rusty iron chute on wooden legs came flying, like a monstrous gargoyle, across the parapet.

—Stevenson, *Silverado*, 1883

There is a closeness to everything: the shiny stalactites behind you blending into the darkness, the dripping of water, the dense shadows of the stalagmites. "Beautiful," you whisper.

You are surrounded on all sides by deep galleries of quartz rising to support the cavern roof. You gaze up at tiny amber stalactites that cover much of the roof, beautifully arched. Tears almost come to your eyes.

Theano smiles and nods. "Incredible," she says.

You feel goosebumps rise along your arms as you look at Theano. "I, I wish I had met you sooner, before the transfinites appeared, before I became preoccupied with the End of the World, before I knew a comet would someday hit the Earth." You stare into her lovely eyes which bioluminesce like fireflies in the shadows of the stalactites.

The silvery light from the flashlight is falling on her brown hair. Her eyes are moist. How beautiful, yet sad, she seems.

"We're together now," she says as she presses her hand into yours. She hugs you, running her fingers up through your hair.

Mr. Plex coughs. "Sir, what is our lesson today?"

You take a deep breath. "Gargoyles."

Historical gargoyles from the Cathedral of Notre Dame, Paris.

As Mr. Plex squats down to listen, his joints make a peculiar crunching sound like the breaking of bones.

You reach behind a stalagmite and withdraw a stone creature resembling a devil. "Got this from the Cathedral of Notre Dame in Paris."

Theano gasps.

You run your hand along the rough stone. "The term *gargoyle* originally referred to grotesque water spouts shaped like bizarre animals or monsters. They projected from the gutters of buildings, especially in Gothic architecture, carrying the rain-water clear of the building. Today they also refer to grotesque masks and other ghostlike visages."

"Sir, what's this have to do with mathematics?"

"We're going to talk about numerical gargoyles. They're based on a cellular automaton consisting of a grid of cells which can exist in two states, occupied or unoccupied. The occupancy of a cell is determined from a simple mathematical analysis of the occupancy of neighbor cells." You pause as you stretch your arms. "Though the rules are simple, the patterns are very complicated and sometimes seem almost random, like a turbulent fluid flow or the output of a cryptographic system."

Numerical gargoyle produced by a mathematical simulation.

"Sir, can we program this on a computer?"

You nod and withdraw a notebook computer from behind a stalagmite. "I've put the program in the machine already. Just turn it on."

Mr. Plex grabs the computer from you, flicks on the switch, and in seconds is mesmerized by the undulating and ghostly physiognomies. "Amazing," he says. "It's hard to believe this comes from such simple rules in a checkerboard world."

You nod. "Look, there's a weird one." You point to a small evolving blob which appears to have two hollow eyes and a screaming mouth. "The space is just teaming with numerical gargoyles. You can use the computer program like a microscope to magnify interesting regions."

"Wild," Theano says.

The reflections from the computer screen make the cave walls pulse like a living lava lamp. A series of pastel circles moves around the walls of the caves, as in a patternless kaleidoscope. Slowly they coalesce into a beautiful blob of bathybius.

"Let's stretch our legs," Theano says.

The three of you get up and walk for a few minutes beneath an inverted forest of bones—ivory stalactites which form the tunnel. A few delicate helicites grow like fungus on the walls.

Theano suddenly cocks her head. "What's that sound?"

You hear wind whistling through the cave and smell a vague scent of ammonia. It feels as if the cave is breathing, inhaling and exhaling great quantities of air.

You turn to Theano. "Nothing to worry about. Just the wind. I don't—" you stop yourself suddenly.

Theano looks into your eyes. "What's wrong?"

You stoop down. "Look at that. A footprint—"

"Gargoyle!" Theano screams with a quick indrawn breath.

"What?" You spin around and see two feeding tubes shooting out from your Notre Dame gargoyle. The tubes slowly trail along the wet cave floor. Red, engorged arteries are braided along the tubes' entire length.

You clench your fists. "Those transfinites. They're trying to scare us again." You turn your face to the ceiling of the cave. "We've done nothing wrong," you say to no one in particular. "We've done our best not to violate the *Main Directive*. We're not going to areas of human habitation anymore. We only want to learn."

Numerical gargoyle produced by a mathematical simulation.

Theano runs toward you. "What do we do now?"

Mr. Plex, as quick and quiet as a ferret, runs up to the gargoyle. He stares at it for a second, then raises up his body like a cobra about to strike, and smashes the gargoyle with his right forelimb.

THE SCIENCE BEHIND THE SCIENCE FICTION

> *"Oh, Gargoyle, darling," she said, sitting down on an old hitching block at the edge of the Rosedale pavement, "isn't it too gorgeous?"*
> —J. Harris, *Weird World Wes Beattie*, 1964

> *The term "gargoylism" has been applied to this syndrome because the gross disfiguration resembles the gargoyles of Gothic architecture.*
> —R. Durham, *Encylcopedia of Medical Syndromes*, 1965

Gargoyles and Gargoylism

Originally the term "gargoyle" referred only to the carved lions of classic cornices or to terracotta spouts, such as those found frequently in Pompeii. The word later became restricted primarily to the grotesque, carved spouts of the Middle Ages. The gargoyle of the Gothic period is usually a grotesque bird or beast. Today, the term "gargoyle" has gradually evolved to include other morbid beasts such as chimeras decorating the parapets of the church of Notre Dame in Paris.

Gargoylism is a syndrome characterized by mental deficiency and skeletal deformities including an abnormally large head, short limbs, and protruding abdomen. The disease is also known has Hurler's syndrome.

Graphical Simulation

Numerical gargoyles undulating through horror movies of the future will include computerized versions where the ghostly faces are mathematical forms displayed on a computer screen. In order to create the monstrous faces, the simulation produces shapes that move, coalesce, and break up on a 2-D grid. The simulation involves the use of "cellular automata." Cellular automata (CA) are a class of simple mathematical systems which are becoming important as models for a variety of physical processes. CA are mathematical idealizations of physical systems in which space and time are discrete. As you told Mr. Plex and Theano, CA usually consist of a grid of cells which can exist in two states, occupied or unoccupied. The occupancy of one cell is determined from a simple mathematical analysis of the occupancy of neighbor cells.

To create a CA, each cell of the array must be in one of the allowed states. The rules that determine how the states of its cells change with time are what determine the CA behavior. There are an infinite number of possible CA, each like a checkerboard world. I produced the shapes for the figures on pp. 182 and 183 by initially filling the CA array with random 1s and 0s. The next section discusses the rules of growth that determine the states of the cells in subsequent generations.

Grow Your Own Gargoyles

The gargoyles in this chapter evolve in discrete time according to a local law. Specifically, the value taken by a cell at time $t + 1$ is determined by the values of its neighbors (and its own value) assumed at time t:

$$c^{t+1} = f(c^t_{(i,j)},\ c^t_{(i+1,j+1)},\ c^t_{(i-1,j-1)},\ c^t_{(i-1,j+1)},\ c^t_{(i+1,j-1)},\ c^t_{(i+1,j)},\ c^t_{(i-1,j)},\ c^t_{(i,j+1)},\ c^t_{(i,j-1)}).$$

In this equation, $c^t_{i,j}$ denotes the state occupied at time t by the site (i,j). The nine-cell template I used to produce the numerical gargoyles is called a *Moore* neighborhood (as opposed to a *von Neumann* neighborhood consisting only of orthogonally adjacent neighbors). For example, let's denote the current cell and its position i (horizontal position) and j (vertical position) in the checkerboard by $c(i,j)$. Its neighbor cells are $c(i+1,j)$, $c(i-1,j)$, $c(i,j+1)$, and $c(i,j-1)$ for the cells up, down, right, and left of the current cell, and $c(i+1,j+1)$, $c(i-1,j-1)$, $c(i+1,j-1)$, and $c(i-1,j+1)$ for the diagonal neighbors. A computer program starts by randomly assigning 1s and 0s to positions in the checkerboard. You can think of this as randomly painting cells white or black. One interesting simulation next examines the neighbor sites to determine if the majority of neighbors are in state one. If so, then the center site also becomes one. We can represent those cells in the on (one) state as black dots on a graphics screen. In other words, this rule is a *voting rule* which assigns 0 or 1 according to the "popularity" of these states in the neighborhood, and interestingly it generates behavior found in real physical systems. (This rule is applied to the newly generated pattern and repeated over and over again to see what patterns evolve as the 1s and 0s change.) This simple *majority rule automata* produces hundreds of coalesced, convex-shaped black areas but does not lead to interesting graphical forms. A way to destabilize the interface between one and zero areas is to modify the rules slightly so that a cell is on if the sum of the one sites in the Moore neighborhood is either 4, 6, 7, 8, or 9, otherwise the site is turned off. (To do this, a computer program simply examines the number of neighbor cells to see how many have a value of 1.) This rule has been studied previously. Since it uses a Moore neighborhood, it was termed *M46789* by Gerard Vichniac in 1986. Such simulations have relevance to percolation and surface-tension studies of liquids.

The gargoyles on pp. 182 and 183 are *M46789* forms that have evolved after a few dozen time steps from random initial conditions on a 500 × 500 square lattice. The Appendix contains BASIC and C program listings for computing numerical gargolyes.

數 數 數 數

In this chapter, we've had a brief introduction to gargoyles. To understand gargoyles, we must imagine the medieval person's strong belief in God and reliance on the Church. The cathedral was a "sermon in stone" which had to be "read" by an illiterate population. Some gargoyles illustrated Bible stories, but most gargoyles were probably meant to scare away evil spirits. Religion, like the gargoyles, existed to confront and overcome chaos and danger.

Gargoyles were often of ambiguous gender and species, and perhaps these images were thought to be particularly frightening. Some have suggested that gargoyles represent pre-Christian practices and symbols incorporated into the rituals of the Catholic Church to facilitate conversions to Christianity.

In addition to these theories, I think one of the best ways to understand gargoyles is to imagine yourself being alive in the Middle Ages, a thousand years ago. Your reality is defined by the Christian Church that says that all of life, from thunderstorms to famines, is either the will of God or the jesting of the devil. Perhaps the gargoyles are a kind of spiritual test. You are placed in the universe to overcome temptations, to choose between the divine and the devil. As the centuries pass, humans become less certain of a God-controlled cosmos, and science starts offering explanations. This shift from a spirit-controlled world to scientific solutions leads to the demise of the gargoyle in the lives of people.

Today we can offer these kinds of rational analyses, but gargoyles were a mystery to many in the past. St. Bernard of Clairvaux wondered in the 12th century:

> What are these fantastic monsters doing in the cloisters under the very eyes of the brothers as they read? What is the meaning of these unclean monkeys, strange savage lions and monsters? To what purpose are here placed these creatures, half beast, half man? I see several bodies with one head and several heads with one body. Here is a quadruped with a serpent's head, there a fish with a quadruped's head, then again an animal half horse, half goat.... Surely if we do not blush for such absurdities we should at least regret what we have spent on them.

CHAPTER 18

Astronomical Computers in Canchal de Mahoma

The future is a fabric of interlacing possibilities, some of which gradually become probabilities, and a few which become inevitabilities, but there are surprises sewn into the warp and the woof, which can tear it apart.
—Anne Rice, *The Witching Hour*

The Stone Age artist ... relied exclusively on his mathematical instinct. It's his instinct that has been abstracted and fettered into geometrical forms. In the course of time, it helped to develop the concept of numbers ... so that finally the manifold manifestations of being in space and time could be put in order by abstract numbers.
—Annemarie Schimmel, *The Mystery of Numbers*, 1993

"Now where have you taken us?"

"Canchal de Mahoma, Spain, 7000 B.C. We're going to see an amazing painting by an Ice Age artist."

You walk through a maze of stalagmites and moist passages winding though one another like worm tunnels.

"Watch out!" Theano screams.

You are slipping feet first into a funnel-like pit which drops off an indeterminable distance. Theano reaches out her hand.

"No," you cry, "I'll take you with me."

"No you won't," Theano screams. "Now grab my hand." Her voice echoes around the cavern like the wails of ghosts.

Mr. Plex reaches his forelimb to you.

A sharp crystal rips through your shirt as you spread your legs in an attempt to stop your descent. "I'm OK," you say holding Mr. Plex's hand. Slowly you manage to climb your way out of the funnel.

Your body shakes. "That was close."

Theano comes closer. "You alright?"

"I'll survive."

Theano nods as she takes deep breaths. "Let's rest," she says.

After a few minutes, your heartbeat returns to normal. The three of you seem relaxed.

Theano's eyes wander over the cave walls. "What's that?" she asks pointing to some marks on the wall. A vertically elongated central smudge is surrounded by spots.

Mr. Plex shuffles closer. "Sir, a war club surrounded by flies?"

You shake your head. "Actually, there *was* an explorer named Abbe Breuil who discovered the pattern and thought the central mark was a war club. But now most archeologists think that the club is really a symbol of God, and the spots are phases of the Moon."

You walk closer to the wall. "Here. Take a look at this. The Moon goes through a complete period of phases in 29 or 30 days. Starting at the bottom of the figure: first there is a crescent curved on the western horizon, then a first quarter resembling half a silver dollar, then full Moon, last quarter, and crescent. There is a period of invisibility between the old and new crescent when the Moon passes in front of the sun."

Theano studies the figure. "Pythagoras called this gap the ενη και νεα, the day when the old changes to the new." She pronounces the three words as "en-i, ke, ne-a."

You nod. "In the cave painting, the lunar sequence starts at the lower left with the bow of the thin crescent facing to the right, exactly as it appears in the sky at sunset. Going counterclockwise, next come six spots, one for each night of the first quarter. The eighth spot is separated, the night of the first quarter. The Moon at the top left corresponds to the end of the month with the crescent facing to the east (left) as it does in the sky."

Astronomical "computer," 7000 B.C., on a cave wall at Canchal de Mahoma, Spain.

"Sir, this seems accurate. The crescents are pointing in the correct directions. The counterclockwise progression is the correct orbital direction of the Moon. The large gap between the old and new crescents corresponds to the days of invisibility, moonless nights."

"Mr. Plex, what's that on your back?" You walk over and tear off a paper stuck to Mr. Plex's exoskeleton with some mucilagenous substance.

Mr. Plex suspiciously eyes the paper in your hand. "Sir, what in heaven's name is it?"

You look at the page. "It seems to be from the Book of Revelation." You pause. "Those transfinites are trying to mess with our minds again."

Mr. Plex makes a sound like a parakeet being vacuumed out of its cage.

Theano is studying the astronomical design on the cave wall. "Pythagoras would've found this diagram fascinating."

You come closer. "Why?"

"On the 17th of the month, it's clear that the period of the full Moon is over. Pythagoras called this day 'the barrier' and hated this number. It was the awful night when the Moon began to die. The number 17 separates the square 16 and the oblong number 18, which are the only rectangular figures that have their areas equal to their perimeters:

$$16 = 4 + 4 + 4 + 4 \text{ and } 16 = 4 \times 4$$
$$18 = 3 + 6 + 3 + 6 \text{ and } 18 = 3 \times 6$$

Pythagoras hated 17 for other reasons, besides it being an indivisible prime number."

"Why?"

"17 keeps the divine 16 apart from its *epogdoon* partner 18."

"Epogdoon?"

"In Greek, 'ogdoos' means 'eighth.' The 'epogdoon' of a number is simply a number that is one-eighth greater than the original. $16 + (1/8) \times 16 = 18$ (its epogdoon)."

The light from the flashlight is dimming.

"Shut it off while we rest," Theano says. "Don't waste the battery."

You turn off the light and look at Theano against the phosphorescent backdrop of the cave. She is surrounded by sparkling stalactites and crystals, a mineralogist's dream. It makes Theano seem all the more alone, lost. Were you right to take her from ancient Greece?

Where is the rest of the world? It hardly matters. You hear the mystical sounds that had lulled you as a boy exploring small caverns—the humming of stalactites, the wild, desperate cry of the wind through the cave, the chilled air.

You click the light back on. Tendrils of violet helicite crystals line a nearby rock. Gypsum chandeliers sparkle like diamonds. "Not a bad place to die," you say, running your fingers along some calcite pearls. You hand a few of the pearls to Theano. "For you, my dear."

Theano snuggles up near a stalagmite, a Greek robe over her shoulders. Her cheeks are slightly flushed. She has never seemed more relaxed.

THE SCIENCE BEHIND THE SCIENCE FICTION

> *For an untouched tribe in the remote Amazon basin, a world run by terrifying humanlike gods and godlike animals provides all the story they need to survive. The myths that must sustain an industrial society in a global village ... must be more useful and complex than for the rain forest tribe, no matter how much that society longs for primitive simplicity.*
> —Wallace Kaufman

There really is the cave Canchal de Mahoma in Spain, and it contains the astronomical diagram pictured in this chapter. The painting was made no later than 7000 B.C.—the end product of art practiced through the Ice Age. Of course, interpretation of the markings is highly speculative.

Pythagoras did call the 17th day of the month "the barrier," and his distaste for the number 17 is well documented by Plutarch. Plutarch also wrote in *Isis and Osiris* that "the dismemberment of the Egyptian god Osiris into 14 parts referred allegorically to the waning days of the moon ... from the time of the full moon to the new moon."

毅 毅 毅 毅

In this chapter, we've discussed an ancient depiction of the Moon's motions, which may have served as a primitive calendar. Various heavenly bodies—for example the Moon, planets, Sun, and other stars— have provided a means for measuring the passage of time since the dawn of humanity. Ancient civilizations relied upon the apparent motion of these bodies through the sky to determine seasons, months, and years. We know little about timekeeping in prehistoric eras, but wherever archeologists find records and artifacts, they also often find evidence that a culture was preoccupied with measuring and recording the passage of time. Ice Age hunters in Europe over 20,000 years ago scratched lines and chiseled holes in bones and sticks, possibly counting the days between phases of the Moon. Five thousand years ago, Sumerians in the Tigris–Euphrates valley in today's Iraq fashioned calendars that divided the year into 30-day months, divided the day into 12 periods (each corresponding to two of our hours), and divided these periods into 30 parts (each like four of our minutes). Stonehenge architects who aligned massive columns of stone over 4000 years ago in England left no written records; however, the alignments of stones and holes suggest that Stonenenge could be used to determine seasonal or celestial events, such as lunar eclipses and solstices.

Other cultures also had early calendars. Egyptians first based their calendars on the lunar cycles, but around 4235 B.C., they used the star Sirius (the Dog Star), which rose next to the Sun every 365 days, to create a 365-day calendar. Some scholars set 4235 B.C. as the earliest recorded year in history.

Before 2000 B.C., the Babylonians (again in Iraq) had a 354-day year based on 12 alternating 29-day and 30-day lunar months. The Mayans of Central America (2000 B.C.– A.D. 1500) relied on the Sun, Moon, and Venus to establish 260-day and 365-day calendars. Their written celestial-cycle records indicated their belief that the world was created in 3113 B.C. Mayan calendars later became portions of the great Aztec calendar stones. Other civilizations, such as our own, use a 365-day solar calendar with a leap year occurring every fourth year.

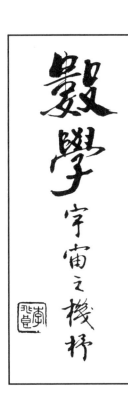

CHAPTER 19

Kabala

"Is there a God, Lasher?"
*"I do not know, Rowan. I have formed an opinion and it is yes, but it fills me
with rage."*
"Why?"
"Because I am in pain, and if there is a God, he made this pain."
"But he makes love, too, if he exists."
"Yes. Love. Love is the source of my pain."

—Anne Rice, *The Witching Hour*

*Geometry is co-eternal with the Mind of God before the creation of things: it
is God Himself.*

—Johannes Kepler (1571–1630)

Theano is taking food from a large box: eggs, a crushed box of matzos, butter, and jelly.
The large box is stamped "Perishable—A.D. 2080."

Mr. Plex crowds against Theano, seizing the matzos with one of his claws, the jelly
with another.

You yawn and walk over to observe Mr. Plex's rapid motions. "Hungry, Mr. Plex?"

Theano breaks a miniature stalagmite from the cave wall and stirs the shivering eggs
in a pan. The three of you sit beside the fire, the smoke rising like an offering from a stone
altar.

You hear music coming from far away. A lone kithara player?

Theano smiles. "I want to show you something. I set it up while you slept." She takes
your hand and leads you to the cave entrance.

There is a large Christmas tree at the mouth of the cave. Some of the stalactites are
decorated with holly. On the ground beneath the tree is a box of Christmas lights and
ornaments.

You turn to Theano. "Magnificent! I didn't know you celebrated such a feast."

"It's an old feast," Theano says. "It goes back centuries before Christ. The winter solstice. An important time astronomically. That's probably why God chose it as a time to be born."

You shake your head. "Where did you get all this?"

Theano smiles at Mr. Plex, who beams with pride. "Mr. Plex helped me," she says.

All around you is the smell of the Christmas tree, sweetly fragrant, and of the fire burning. You feel delicious in the warmth.

What's that you hear? It seems that a kithara is playing somewhere, and a low voice is singing a slow mournful melody, an ancient Greek song. Gradually the song metamorphoses into a Celtic tune about a child lying in a manger.

You turn toward to Theano. "Hear something?"

"What?" Theano is lying on her side, looking at the cave entrance where crusts of frost are forming on the outermost stalactites. Very slowly, a figure begins to take shape—a man with a glowing thigh, facing the cave, his arms folded.

Theano doesn't hear the spirit coming toward the cave. Shadows from the stalactites make the icicle phantom difficult to discern.

Mr. Plex is chewing and sucking on some scrambled eggs. "Sir, what's the lesson for today?"

You shake your head, and the icicle phantom is gone. Perhaps it was merely a fragment from a fading dream.

"Sir?"

"Kabala," you say. "It's an esoteric Jewish mysticism that became popular in the 12th and following centuries. Much of the Old Testament, they claimed, is in code. That's why scripture may seem muddled."

Mr. Plex holds up his forelimb as matzo crumbs fall from his mouth. "Don't beginners need a personal guide to avoid danger?"

You nod. "I'm your guide." You pause. "The early roots of Kabala are traced back to Merkava mysticism."

"Merkava?" Mr. Plex says.

"Merkava is God's throne-chariot as described in Ezekiel 1:26. The goal of early Jewish mystics was to ascend through the heavens and view God's glory (kavod) seated on his throne-chariot." You pause. "In Palestine, Kabala began to flourish in the 1st century A.D. The earliest known Jewish text on magic and mathematics, *Sefer Yetzira* (Book of Cre-

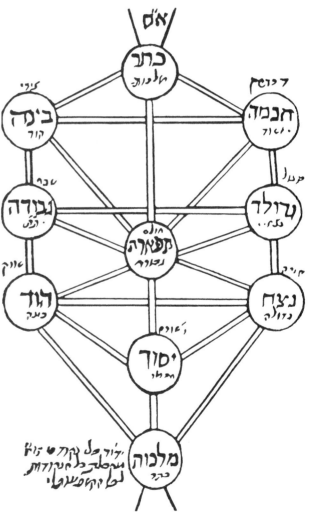

The Sephiroth Tree, or Tree of Life, from an old manuscript of the Zohar.

ation), appeared around the 4th century A.D. It explained creation as a process involving 10 divine numbers or *sephiroth*."

Theano throws some new spruce wood on the fire, and the flames shoot upward.

You warm your hands by the fire. "The Jewish Kabala is the most important development of the Pythagorean tradition in the medieval world." You pause. "Just like the Pythagoreans, the Jewish Kabalists considered numbers sacred and had a particular interest in the number 10."

Mr. Plex jumps up. "Mon Dieu! The tetraktys. There's that number 10 again."

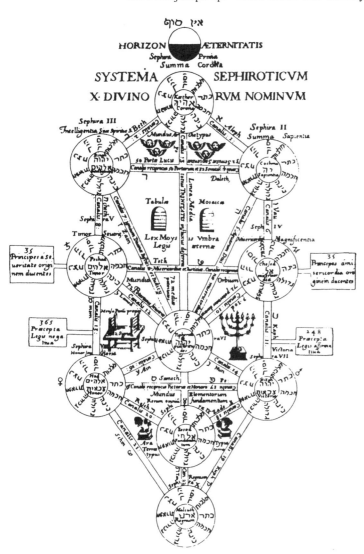

Another representation of the sephiroth, the central figure for the Kabala.

You nod. "Kabala is based on a complicated number mysticism whereby the primordial One divides itself into 10 sephiroth which are mysteriously connected with each other and work together. 22 letters of the Hebrew alphabet are bridges between them."

You grab a piece of charcoal and begin to sketch on the cave wall. "The highest sephiroth is *keter* (crown) out of which *hokhmah* (wisdom) and *binah* (intelligence) branch. The other sephiroth are: love, greatness, justice, beauty, triumph, splendor, fundament, and finally *malkhut* (or Kingdom or Reality). This last sepheria can be equated with the *Shekhinah* that live in the exile of this world."

You begin to draw lines on your diagram. "Like I said, there are 22 ways that the sephiroth are connected. The Sephiroth are 10 hypostatized attributes or emanations allowing the infinite to meet the finite. (*Hypostatize* means to make into or treat as a substance, to make an abstract thing a material thing.) Through study of the 10 sephiroth and their interconnections, one can develop the entire divine cosmic structure."

Mr. Plex throws another branch onto the fire. "Sounds heavy."

Theano stretches her legs. "For my husband, 10 played a magical role. Was it magical for the Jews in other areas?"

You nod. "Ten appears often in Judaism." You begin to outline your thoughts on the
cave wall as you speak:

193

Kabala

- There are 10 commandments.
- The Zohar, the central text of the Kabala, says the world was created in 10 words, because in Genesis 1, the phrase "And God spoke" is repeated not less than 10 times.
- There are 10 generations between Adam and Noah.
- There are 10 plagues in Egypt.
- On 10 Tishri, the Jewish Day of Atonement, the confession of sins is repeated 10 times.
- On Rosh Hashanah, the Jewish New Year, 10 biblical verses are read in groups of 10.

Mr. Plex has finished the box of matzos and is chewing on the cardboard box. "Sir, did the Kabalists use the sephiroth to interpret the Bible?"

You rip the matzo box from Mr. Plex's forelimbs. "Yes. They gave each Hebrew letter a numerical value, a method known as *gematria*. For example, consider the passage in Genesis (49:10): 'The scepter shall not depart from Judah, nor a lawgiver from between his feet, until Shiloh come; and unto him shall the gathering of the people be.' " You pause. "In Hebrew, 'until Shiloh come' is *IBA ShILH* (*Yavo Shilo*), and if we add together the numerical values of each letter, we get a total of 358. Since the Hebrew for messiah is *MshIch* (*Meshiach*), and the letters also total 358, the Kabalists interpret this phrase from Genesis as a prophecy of the coming of the Messiah."

You gaze into the burning fire. The only other illumination comes from the tree, which Theano has strung with countless tiny, twinkling lights.

The prophecy of the coming of the Messiah. Judgment day. End of the world. Why have we always been concerned with how the world dies? You sigh. Soon you will view the end of the Earth. You, too, are fascinated by endings.

You turn to Mr. Plex and Theano. "Kabalists also used numbers to understand the *tetragrammaton* or four-letter name of God. They wrote the name as *IHVH* which has become Jahweh or Jehova in English, but it was originally considered so sacred that it shouldn't be pronounced. In fact, the only reason the Jews remembered it was that elders were allowed to pass it on to their disciples once every seven years."

Theano's eyes are clear, direct, penetrating. "This sort of thing would have made Pythagoras happier than a Centaur in heat."

"Wait," you say. "The best is yet to come. According to the Kabalists, Jahweh is itself a substitute for the true name of God which had 72 syllables and 216 letters! They call this holy name the *Shem ha-meforash*."

"Mon Dieu!"

"Wait! To create the *Shem ha-meforash*, go to verses 19–21 of Exodus 14. Verse 19, 20, and 21 each have 72 Hebrew letters. Write Verse 19. Beneath it write Verse 20 in reverse order. Beneath Verse 20 write Verse 21. If you read from top down this creates 72 3-letter names, all of which can be connected together to make one name of God. Also, add to these 3-letter names AL or IH, and this creates 72 names for 72 different angels."

"Wild," Theano whispers.

"Wait! Amazingly, if the letters of *IHVH* are arranged in the form of a tetraktys, assigned their numerical equivalents, and added together line by line, the total comes to 72. Kabalists use the *Shem ha-meforash* and the 72 angel names to coax God and his angels to answer prayers."

Theano's eyes are unfocused. "All the numbers are swimming in my head," she says. You take a breath. "Me, too."

Theano stands watching you as you climb a stalagmite and make some little adjustment to the tree. She is whistling a soft kithara song. So mournful, it makes you think of a deep ancient wood in winter. She leans against a stalagmite and looks at the immense tree all speckled with its tiny lights like stars, and breathes its deep woody perfume.

Theano walks over to a box of ornaments. "Aren't they lovely!" She picks up a small white angel with golden hair. She replaces it and picks up a golden infinity symbol, with the faintest blush to its curves. "They're beautiful. Where do they come from?" She lifts a silver fractal Koch curve.

Your recognize the box Mr. Plex has retrieved from the future. "I've had them for years. I could never have imagined that I would be using them to decorate a tree in a cave with you back in time." You pause. "Pick one for the tree."

"Koch curve star," she says.

Theano hands the silver Koch curve to you, its triangles upon triangles glistening like diamonds. For an instant you feel like light trapped within a tube, totally internally reflected. You peer into the snowflake curve, your image many times reflected as if you are standing on the periphery of some gigantic crystal, alone in a field of darkness.

You lift the Koch snowflake and place it on a branch. The fractal snowflake shivers. You are mesmerized by the lovely play of light on the deep green fractal branches. You need not move. There is enough motion from the lights. You feel like you can live forever, suspended in space.

The light is reflected according to mathematical laws. Angles, polarizations, intensities, refractions, diffractions, interferences, geometrical optics, spherical aberrations. Such beauty from pure math.

Theano snaps you from your reverie. "Here, I've gotten something for you," she says. "Merry Christmas." She places in your hands a small vessel wrapped in red Christmas paper. "It's not a very big present, I'm afraid, but it's the best I could find."

You quickly unwrap the gift and smile. "Kylonian brandy, my favorite."

Theano smiles. "Mr. Plex helped retrieve it for me from the past."

You turn to Mr. Plex. "You are a good friend."

Suddenly, there is a vague humming sound coming from deeper in the cave, like a church choir tuned two octaves lower.

"What's that?" Theano says.

"Let's check it out."

You, Theano, and Mr. Plex walk a little further and see a small subchamber of the cave.

Mr. Plex stops. "Shall we go in?"

You nod. "Let's find out all we can."

Entering quietly, the three of you stand still and survey a bizarre, cold world of emaciated sleepers against the cavern walls.

"Oh, my God," Theano whispers. "Th–there's hundreds of them."

Each alien appears as if it is in suspended animation. Their faces are partially covered by some kind of breathing mask with a snaking tube that goes into a hole in the wall. Their fragile limbs are hooked to flashing monitors amid a maze of cables and gauges.

Mr. Plex's forelimbs are shaking. "Is it—?"

You nod. "The transfinites. What they really look like."

A loud booming sound rolls across the cave, but you hardly notice the ominous beating. To you, the pounding of your own heart is the loudest noise in the cave.

A scream seems to be swelling inside Theano. As she holds it back you can only imagine the unreleased pressure as a painful burning in her chest.

The alien nearest you has skin that is bleached whiter than bone. It stands there against the wall, seemingly lifeless, its insectile head encased in an enormous turban of wires and tape. A thick amber tube runs into its mouth. A thin needle is taped to its right forelimb. The only sign of life is the blinking of the machines—machines that probably keep the creatures alive or quiet in this dormant state.

You back up. "Let's get the hell out of here," you say. The creatures' gelatinous orbs sparkle like jewels. Their eyes are staring at you and Theano.

"Wait," you say.

You feel like walking over to the support machinery, finding a power switch, and shutting it all off. But you can't bring yourself to kill hundreds of sleeping creatures no matter how much trouble they've caused.

THE SCIENCE BEHIND THE SCIENCE FICTION

Kabala was originally the post-biblical Hebrew name for the oral tradition handed down from Moses to the rabbis of the Mishnah and the Talmud. Later, toward the beginning of the 13th century, the term was applied to the mystical, numerical interpretation of the Old Testament. Usually a personal guide was used so initiates could avoid personal danger when unlocking the mysteries of the universe.

A major text of early Kabala was the *Sefer ha-bahir* (Book of Brightness, 12th century), which not only interpreted the sephiroth as essential in creating and sustaining the universe but also introduced into Judaism the concept of transmigration of souls (*gilgul*).

Following their expulsion from Spain in 1492, the Jews became increasingly interested in eschatology and messiahs, and the Kabala became more popular. By the mid-16th century, Safed, Galilee, was the center of the Kabala. In this sacred locale, the greatest of all Kabalists, Isaac ben Solomon Luria, spent the last years of his life. Lurianic Kabala had several main tenets:

- The withdrawal (*tzimtzum*) of the divine light, thereby creating primordial space.
- The sinking of luminous particles into matter (*quellipot*).

● Cosmic restoration (*tiqqun*) achieved by the Jew through an intense mystical life and struggle against evil.

Lurianic Kabbalism was used to justify *Shabbetaianism*, a Jewish messianic movement of the 17th century. Lurianic Kabbalism also influenced the doctrines of modern Hasidism, a religious and social movement started in the 18th century and still flourishing today in certain Jewish communities.

As mentioned in Chapter 5 the number 7 plays an important role for Kabalists. In the 12th Psalm, it is said that the words of Yaheh are pure words, purified 7 times. The Kabalists use this fact to better understand biblical stories. The 7 lower sephiroth correspond to the Temple. The 7th of these lower sephiroth is the Shekhinah, which is called the Sabbath Queen, corresponding to the 7th primordial day.

The Kabalistic Loom

In the Kabalistic world view, everything is symbolic for something else, and the world is filled with subtle connections, as if reality were constructed from invisible threads binding events and ideas. Through study of the 10 sephiroth and their interweavings, one seeks to understand the cosmic structure. According to Kabalists, even the most trivial text, number, or object hides deep secrets. Kabalists believe that God created the universe by allowing some of his quintessence to flow "down," transmuting into the material universe. The goal for many Kabalists is to understand this process, the universe, and the way back to God. The concept of God is complex. God has names with different powers ruling over hierarchies of angels, but God is also a completely abstract entity. Many Kabalists believe it is impossible to understand the nature of God, except to understand possibly what things God is not.[1]

A Loom of Universal Strings

The sephiroth as 10 hypostatized attributes linking the infinite and the finite remind me that physical reality may be the hypostatization of mathematical constructs called "strings." Strings, the basic building blocks of nature, and not tiny particles but unimaginably small loops and snippets of what loosely resembles string—except that the string exists in a strange, 10-dimensional universe. The current version of the theory took shape in the late 1960s.

In the last few years, theoretical physicists have been using strings to explain all the forces of nature—from atomic to gravitational. As mentioned, string theory describes elementary particles as vibrational modes of infinitesimal strings that exist in 10 dimensions. How could such things exist in our four-dimensional space–time universe? String theorists claim that six of the ten dimensions are "compactified"—tightly curled up (in structures known as Calabi-Yau spaces) so that the extra dimensions are essentially invisible. Unfortunately, there are so many different ways to create universes by compactifying the six dimensions that string theory is difficult to relate to the real universe. In

1995, researchers suggested that if string theory takes into account the quantum effects of charged mini black holes, the thousands of four-dimensional solutions may collapse to only one. Tiny black holes, with no more mass than an elementary particle, and strings may be two descriptions of the same object. Thanks to the theory of mini black holes, physicists now hope to follow the evolution of the universe mathematically and select one particular Calabi-Yau compactification—a first step to a testable "theory of everything."

Just like the early years of Einstein's theory of relativity, string theory is simply a set of clever equations waiting for experimental verification. Unfortunately, it would take an atom smasher thousands of times as powerful as any on Earth to test the current version of string theory directly. I hope humans will refine the theory to the point where it can be tested in real-world experiments. Edward Witten, a professor at the Institute for Advanced Study in Princeton, has worked on a related idea known as topological quantum field theory, which allows physicists to find connections between seemingly unrelated equations. With Witten directing his attention to string theory, it is hoped that he and his colleagues can crack the philosophical mystery that's dogged science ever since the ancient Greeks: What is the ultimate nature of the universe? What is the loom upon which God weaves?

Whatever the loom is, it has created a structurally rich universe. Most astronomers today believe that the universe is between 8 billion and 25 billion years old, and has been expanding outward ever since. The universe seems to have a fractal nature with galaxies hanging together in clusters. These clusters form larger clusters (clusters of clusters). "Superclusters" are clusters of these clusters-of-clusters. In recent years, there have been other baffling theories and discoveries. Here are just a few:

- In our universe exists a *Great Wall* consisting of a huge concentration of galaxies stretching across 500 million light-years of space.
- In our universe exists a *Great Attractor*, a mysterious mass pulling much of the local universe toward the constellations Hydra and Centaurus.
- There are *Great Voids* in our universe. These are regions of space where few galaxies can be found.
- *Inflation theory* continues to be an important theory describing the evolution of our universe. Inflation theory suggests that the universe expanded like a drunken balloon-blower's balloon while the universe was in its first second of life.
- The existence of *dark matter* also continues to be hypothesized. Dark matter consists of subatomic particles that may account for most of the universe's mass. We don't know of what dark matter is composed, but theories include: neutrinos (subatomic particles), WIMPs (weakly interacting massive particles), MACHOs (massive compact halo objects), or black holes.
- *Cosmic strings and cosmic textures* are hypothetical entities which distort the space-time fabric.

What are the biggest questions? Perhaps, "Which laws of physics are fundamental and which are accidents of the evolution of this particular universe?" Or, "Does intelligent, technologically advanced life exist outside our Solar System?" and "What is the nature of consciousness?"

How many of these questions will we ever answer?

CHAPTER 20

Mathematical Proofs of God's Existence

The God of the Old Testament is a God of power, the God of the New Testament is a God of love; but the God of the theologians, from Aristotle to Calvin, is one whose appeal is intellectual: His existence solves certain puzzles which otherwise would create argumentative difficulties in the understanding of the universe.

—Bertrand Russell, *A History of Western Philosophy*

It is in a way a forlorn and perhaps even hopeless objective—to demonstrate the existence of God by numerical coincidences to an uninterested, to say nothing of a mathematically unenlightened public.

—Carl Sagan, *Broca's Brain*

Euler strode up to Diderot and proclaimed: "Monsieur, $(a + b^n)/n = X$, donc Dieu existe!" *["Sir $(a + b^n)/n = X$, therefore God exists!"]*

—Michael Guillen, *Bridges to Infinity*

The cave is bright in the morning light. A few drops of cool liquid fall from a stalactite onto your outstretched palm. You turn toward Theano and Mr. Plex who are sitting with their backs against the cave wall.

"Today, I want to tell you about the use of computers, mathematics, and logic to prove the existence of God."

Mr. Plex's eyes seem to light up like little lanterns. "Is that possible, Sir?"

"Many have tried. Several 20th-century proofs involved mathematics, but some of the most famous proofs of God's existence started with Saint Thomas Aquinas. He lived from 1225 to 1274 and gave five proofs of God's existence in his *Summa Theologica*. The flavor of many of his arguments is as follows. In this universe, there are things which are only moved, and other things which both move and are moved. Whatever is moved is moved by something, and since an endless regress is impossible, we must arrive somewhere at something which moves without being moved. This unmoved mover is God."

Mr. Plex scratches his head. "Not very persuasive."

You nod. "While this argument may not seem very persuasive today, during Aquinas' time it caused quite a stir. Interestingly, Aquinas loved to make lists of things which God cannot do. He cannot be a body or change Himself. He cannot fail. He cannot forget or grow tired, or repent or be angry or sad. He cannot make a man without a soul or make the sum of the angles of a triangle less than 180 degrees. He cannot make another God, cannot kill Himself, and cannot undo the past or commit sins."

Mr. Plex shifts his position against the cave wall as a drop of clear water falls upon his forelimb. "Sir, what about Aristotle?"

"Both Aquinas and Aristotle didn't like an infinite regression of causes, and they used their dislike to demonstrate God's existence. But these great minds lived before infinite mathematical series were commonplace concepts. How would the history of religion have changed if Aristotle and Aquinas knew integral calculus and infinite sequences? Would Aquinas have been better prepared to conceive of an infinitely old universe requiring no creator?" You pause. "Carl Sagan, a great 20th-century science popularizer, often talked about Aquinas' and Aristotle's unmoved mover God. I recall from Sagan's book *Broca's Brain*, 'As we learn more and more about the universe, there seems less and less for God to do. Aristotle's view was of God as an unmoved prime mover, a do-nothing king who establishes the universe and then sits back and watches the intertwined chains of causality course down through the ages.' "

Theano is stirring a pool of water with her finger. The ripples make faint splashing sounds as they contact a central stalagmite. She looks at you. "Sounds like lots of people have tried to prove God's existence using mathematical proofs, but has anyone tried to *disprove* the existence of God?"

"Yes. The early Christian writer Lucius Lactantius quotes the Greek atheist Epicurus in *The Anger of God*:

> God either wishes to take away evil, and is unable, or He is able, and is unwilling; or He is neither willing nor able, or He is both willing and able. If He is willing and is unable, He is feeble, which is not in accordance with the character of God. If He is able and unwilling, He is envious, which is equally at variance with God. If He is neither willing nor able, He is both envious and feeble, and therefore not God. If He is both willing and able, which alone is suitable to God, from what source then is evil? Or why does He not remove evil?

The wind grows stronger near the mouth of the cave, and the blue of the sky outside the cave was fast turning to beige. The dwindling light tinged the cave the color of salmon. Occasionally, you hear strange animal cries and the sounds of insects. You look at Theano who appears to have a slight shiver.

"I'm OK," she says.

You continue. "In one of his more skeptical moods, St. Thomas Aquinas wrote in *Summa Theologica*:

> It seems that God does not exist; because if one of two contraries be infinite, the other would be altogether destroyed. But the name God means that He is infinite goodness. If,

therefore, God existed there would be no evil discoverable; but there is evil in the world. Therefore God does not exist. Further, it is superfluous to suppose that what can be accounted for by a few principles has been produced by many. But it seems that everything we see in the world can be accounted for by other principles, supposing God did not exist. For all natural things can be reduced to one principle, which is human reason, or will. Therefore there is no need to suppose God's existence.

Theano stares intently at you. "What do you think? Can mathematics be used to prove God exists?"

"I don't believe that mathematical 'proofs' can be used to prove or disprove the existence of God, but ever since the time of Pythagoras, philosophers have attempted just this. Each philosopher in turn has found fault with his predecessor: Saint Thomas rejected Saint Anselm's proofs, and Kant rejected Descartes ..."

"Sir, what about some more recent examples?" says Mr. Plex.

"Perhaps the most interesting example of a mathematician studying cosmic questions is Austrian mathematician Kurt Goedel who lived from 1906 to 1978. Sometime in 1970, Goedel's mathematical proof of the existence of God began to circulate among his colleagues. The proof was less than a page long and caused quite a stir. Also, a German mathematician Georg Cantor, who lived from 1845 to 1918, was interested in the way mathematics may imply the existence of God. In letters to Cardinal Franzelin, Cantor explicitly indicated that the infinite, or Absolute, belonged uniquely to God. Cantor developed a mathematical theory of different levels of infinity inhabited by *transfinite numbers*. These are infinite numbers, which we symbolize by the Hebrew letter aleph." You draw the $\aleph$ symbol on the cave floor. "These numbers are used today in mathematics dealing with sets of numbers. For example, the smallest transfinite number is called 'aleph-nought.'" You sketch $\aleph_0$ on the cave floor. "This number counts the number of integers. There are even larger infinities, such as the number of irrational numbers like the square root of 2 which cannot be expressed as a fraction. Cantor believed that God ensured the existence of these transfinite numbers. Cantor regarded the transfinite numbers as leading directly to the Absolute, to the one 'true infinity' whose magnitude was capable of neither increase nor decrease but could only be described as an absolute maximum that was incomprehensible within the bounds of human understanding. The absolute infinite was beyond determination, since once determined, the Absolute could no longer be regarded as infinite, but was necessarily finite by definition."

Theano jumps back from the pool of water. "What are those?"

You come closer, and see numerous flesh-colored objects moving in the pool. "Ah, just some blind cave fish. Amazing how they evolved to function so well without eyes. Their non-visual senses must be very acute to compensate for lack of vision."

Mr. Plex looks at the fish, and then taps you on the shoulder. "Sir, what did Cantor's colleagues think about his ideas on God and infinity?"

"Good question, Mr. Plex. Constantin Gutberlet, one of Cantor's contemporaries, worried that Cantor's work with mathematical infinity challenged the unique, 'absolute infinity' of God's existence. However, Cantor assured Gutberlet that instead of diminishing the extent of God's dominion, the transfinite numbers actually made it greater. After talking to Gutberlet, Cantor became even more interested in the theological aspects of his own theory on transfinite numbers."

Theano looks up from the pool of blind cave fish. "Seems like Gutberlet should like Cantor's ideas because they make God's universe all the more impressive."

"Right. Gutberlet subsequently made use of Cantor's ideas and claimed God ensured the existence of Cantor's transfinite numbers. God also ensured the ideal existence of: infinite decimals, the irrational numbers, and the exact value of π. Gutberlet also believed that God was capable of resolving various paradoxes which seem to arise in mathematics. Furthermore, Gutberlert argued that since the mind of God was unchanging, then the collection of divine thoughts must comprise an absolute, infinite, complete closed set. Gutberlet offered this as direct evidence for the reality of concepts like Cantor's transfinite numbers."

You turn your attention from the fish to Mr. Plex and Theano. "Cantor's own religiosity grew as a result of his contact with various Catholic theologians. In 1884, Cantor wrote to Swedish mathematician Gösta Mittag-Leffler explaining that he was not the creator of his new work, but merely a reporter. God had provided the inspiration, leaving Cantor only responsible for the way in which his papers were written, for the organization and style, but not for their content. Cantor claimed and believed in the absolute truth of his 'theories' because they had been *revealed* to him. Thus, Cantor saw himself as God's messenger, and he desired to use mathematics to serve the Christian Church."

You throw a tiny pebble into the pool, and even before it hits, the fish seem to swim away. "Like Pythagoras, Cantor also believed that numbers (particularly his transfinite numbers) were externally existing realities in the mind of God. They followed God-given laws, and Cantor believed it was possible to argue their existence based on God's perfection and power. In fact, Cantor said that it would have diminished God's power had God *only* created finite numbers. On the other hand, Cantor's love of the infinite had a distinctly anti-Pythagorean flavor. Pythagoras believed infinity was the destroyer in the universe, the malevolent annihilator of worlds. If mathematics were war, the struggle was between the finite and infinite. The Pythagoreans became obsessed with infinity, and they concluded that numbers closest to one (and finiteness) were the most pure. Numbers beyond the range of ten were further from one and were less important. Cantor would not have agreed."

"Sir, it would be intriguing to gather Pythagoras, Cantor, and Goedel in a small room with a single blackboard to debate their various ideas on mathematics and God."

"You bet! What profound knowledge might we gain if we had the power to bring together great thinkers of various ages for a conference on God and mathematics? Would a round-table discussion with Pythagoras, Cantor, and Goedel produce less interesting ideas than one with Newton and Einstein? Could ancient mathematicians contribute *any* useful ideas to modern mathematicians? Would a meeting of time-traveling *mathematicians* offer more to humanity than other scientists, for example biologists or sociologists?" You pause. "These are all fascinating questions. I don't have the answers."

The three of you stare at the school of fish and watch them move in synchrony, despite their lack of eyes. The resulting patterns are hypnotic, like the reflections from a hundred pieces of broken glass. You imagine that the senses place a filter on how much humans can perceive of the mathematical fabric of the universe. If the universe is a mathematical carpet, then all creatures are looking at it through imperfect glasses. How might humanity perfect those glasses? Through drugs, surgery, or electrical stimulation of the brain? Probably our best chance is through the use of computers.

"Proofs" of God in Science Fiction

Science fiction, like science, is an organized system that, for many, takes the place of religion in the modern world by attempting a complete explanation of the universe. It asks the questions—where did we come from? why are we here? where do we go from here?—that religions exist to answer. That is why religious science fiction is a contradiction in terms although science fiction about religion is commonplace.
—James Gunn, *The New Encyclopedia of Science Fiction*, 1988

The brain, knowing that a person can't live forever, rationalizes a future, other-dimensional world in which immortality is possible.

—Philip Jose Farmer

Perhaps the most famous computer proofs of God's existence come from the realm of science fiction. One favorite example is Arthur C. Clarke's "The Nine Billion Names of God" (1953). In this story, Tibetan monks install a computer to calculate and list all possible names of God. Western computer technicians soon learn that when the monks complete their computer explorations, humanity's reason for existence will end. Unfortunately, the technicians learn of the computer too late—just as the project reaches its goal. The technicians look up into the sky, and, "Overhead, without any fuss, the stars were going out." Interestingly, in Clarke's next short story "The Star" (1955), a Jesuit scientist comes to the conclusion that the Star of Bethlehem was actually a supernova which simultaneously guided the wise men to Jesus and destroyed an advanced civilization. The Jesuit struggles to understand why a supernova which annihilated a beautiful, harmless civilization was also the harbinger of hope and salvation on Earth.

In Frederic Brown's "Answer" (1954), computers on many planets are linked together and asked, "Does God exist?" After some time, the computers answer: "He does now."

There are many examples where religion and science fiction intersect. Amazingly, one science-fiction writer, L. Ron Hubbard, actually *created* a religion called Scientology which is practiced by many followers today on Earth. Scientology evolved from dianetics, a method of introspection and discussion invented by Hubbard. Using dianetics, followers attempt to attain mental and physical health. Hubbard first published the theory in *Astounding Science Fiction* magazine in 1950. In the mid-1950s, Scientology took on more religious tenets including the concept of the Thetan, an internal spirit which is endlessly reincarnated.

Here are some additional examples. In one famous science fiction novel, *Canticle for Leibowitz* (1959), Walter Miller describes how post-nuclear-holocaust monasteries collect blueprints and technological artifacts in the same way that medieval monasteries preserved classical manuscripts. Olaf Stapeldon in his book *The Star Maker* (1937) suggests that the stars are sentient and responsible for life on their planets. (Interestingly, *The Star Maker* was not published as science fiction.) In L. P. Gratacap's *The Certainty of a Future Life in Mars* (1903), Mars is the location of Heaven and is stuffed with souls. In John Jacob Astor's *A Journey in Other Worlds* (1894), spirits are found on Saturn.

Doomsday is a popular theme for science fiction writers. For example, in Camille Flammarion's *The End of the World* (1893), the pope and all the clergy pray for the world to be saved from an approaching meteor, but the meteor hits the Vatican, and the book hints that this was God's intention.

In Nils Parling's *The Cross* (1957), a born-again preacher drags a giant, American, atomic cannon around Europe after the Third World War. The cannon, which he calls the fist of God, is thought to be unloaded. Unfortunately, this is not true, and the cannon fires, bringing about an end to higher civilization. The author suggests this is God's way of returning humanity to an ignorant state.

Philip K. Dick wrote about God in *Our Friends from Frolix 8* (1970). The lines I recall most vividly are, "God is dead. They found his carcass in 2019. Floating out in space near Alpha."

I have also touched on the topic of God and mathematics in my own science fiction. For example, in *Chaos in Wonderland* there is a religion devoted to the fractal Mandelbrot set, an intricate mathematical object. (See Chapter 13 for more information on fractals.) Mandelbrot sets also appear on mezuzahs hung on doors and in chapels. I also collaborated with Don Webb on a short story titled "To the Valley of the Sea Horses" that appears in *Keys to Infinity*. The story describes a billionaire's mystical encounter with love, God, and fractals.

Background on Kurt Goedel

> *Theology is a branch of physics.... Physicists can infer by calculation the existence of God and the likelihood of the resurrection of the dead to eternal life in exactly the same way as physicists calculate the properties of the electron.*
> —Frank Tipler, *The Physics of Immortality*

> *A friend of mine once was so struck by [a recursive plot's] infinitely many infinities that he called it "a picture of God," which I don't think is blasphemous at all.*
> —Douglas Hofstadter, *Goedel Escher Bach*

In this chapter, we briefly mentioned historical examples of the co-mingling of religion and mathematics. Perhaps the most interesting example of a mathematician studying cosmic questions is Austrian mathematician Kurt Goedel (1906–1978) (also spelled Gödel). Sometime in 1970, Goedel's mathematical proof of the existence of God began to circulate among his colleagues. The proof was less than a page long and caused quite a stir. Before presenting the essence of his proof in Postscript 1, I want to tell you a little about Goedel. His academic credits were impressive. For example, he was a respected mathematician and a member of the faculty of the University of Vienna starting in 1930. He also was a member of the Institute of Advanced Study in Princeton, New Jersey. He emigrated to the United States in 1940.

Goedel is most famous for his theorem that demonstrated there must be true formulas in mathematics and logic that are neither provable nor disprovable, thus making mathematics essentially incomplete. (This theorem was first published in 1931 in *Monatshefte für Mathematik und Physick* (volume 38).) Goedel's theorem had quite a sobering effect upon logicians and philosophers because it implies that within any rigidly logical mathematical system there are propositions or questions that cannot be proved or disproved on the basis of axioms within that system, and therefore it is possible for basic axioms of arithmetic to give rise to contradictions. The repercussions of this fact continue to be felt and debated. Moreover, Goedel's article in 1931 put an end to a century's long attempt to establish axioms that would provide a rigorous basis for all of mathematics.

Over the span of his life, Goedel kept voluminous notes on his mathematical ideas. Some of his work is so complex that mathematicians believe many decades will be required to decipher all of it. Author Hao Wang writes on this very subject in his book *Reflections on Kurt Goedel* (1987):

> The impact of Goedel's scientific ideas and philosophical speculations has been increasing, and the value of their potential implications may continue to increase. It may take *hundreds of years* for the appearance of more definite confirmations or refutations of some of his larger conjectures.

Goedel himself spoke of the need for a physical organ in our bodies to handle abstract theories. He also suggested that philosophy will evolve into an exact theory "within the next hundred years or even sooner." He even believed that humans will eventually disprove propositions such as "there is no mind separate from matter."

Background on Cantor

> *Cantor was careful to stress that despite the actual infinite nature of the universe, and the reasonableness of his conjecture that corporeal and aetherical monads were related to each other as powers equivalent to transfinite cardinals $\aleph_0$ and $\aleph_1$, this did not mean that God necessarily had to create worlds in this way.*
> —Joseph Dauben, *Georg Cantor*, 1979

> *If the universal resurrection is accomplished by reassembling the original atoms which made up the dead, would it not be logically impossible for God to resurrect cannibals? Every one of their atoms belongs to someone else.*
> —Frank Tipler, *The Physics of Immortality*

Goedel was not the only mathematician to use mathematics to prove or imply the existence of God. As a second example, let us consider Georg Cantor (1845–1918), a German mathematician who also delved into the arena of theomatics.

Cantor was interested in different levels of infinity and in *transfinite numbers*. A transfinite number is an infinite cardinal or ordinal number. (A cardinal number is a whole number, an integer, used to specify how many elements there are in a set. An ordinal number is considered as a place in the ordered sequence of whole numbers. For example, it is used in counting as first, second, third, fourth, etc., to nth in a set of n elements.) The smallest transfinite number is called "aleph-nought" (written as $\aleph_0$), which counts the number of integers. If the number of integers is infinite ($\aleph_0$), are there yet higher levels of infinity? It turns out that even though there are an infinite number of integers, rational numbers (numbers which can be expressed as fractions), and irrational numbers (like $\sqrt{2} = 1.41 \ldots$ which cannot be expressed as a fraction), the infinite number of irrationals is in some sense greater than the infinite number of rationals and integers. To denote this difference, mathematicians refer to the infinity of rationals or integers as $\aleph_0$ and the infinite number of irrationals as C which stands for the cardinality of the "continuum." There is a simple relationship between C and $\aleph_0$. It is $C = 2^{\aleph_0}$. The "continuum hypothesis" states that $C = \aleph_1$;

however, the question of whether or not C truly equals $\aleph_1$ is considered undecidable. In other words, great mathematicians such as Kurt Goedel proved that the hypothesis was a consistent assumption in one branch of mathematics. However, another mathematician Paul Cohen proved that it was also consistent to assume the continuum hypothesis is false!

Interestingly, the number of rational numbers is the same as the number of integers. The number of irrationals is the same as the number of real numbers. (Mathematicians usually use the term "cardinality" when talking about the "number" of infinite numbers. For example, true mathematicians would say that the "cardinality" of the irrationals is known as the continuum.)

What do we do with the paradox of the continuum hypothesis? Cantor's colleague, Constantin Gutberlet, believed that God could resolve the problem of the continuum hypothesis. How many of the great mathematical paradoxes would melt away if humanity had a higher level of intelligence? How many would remain because they are somehow part of the mathematical tapestry underpinning our universe? These are questions not easily answered, at least not by *Homo sapiens*. Our minds have not sufficiently evolved to comprehend all the mysteries of God and mathematics.

A dog cannot understand Fourier transforms or gravitational wave theory. Human forebrains are a few ounces bigger than a dog's, and we can ask many more questions than a dog. Linguist Noam Chomsky once noted that a rat can learn to turn left at every second fork in a maze, but not at every fork corresponding to a prime number. The human mind, limited by the same kinds of biological constraints as the rat, may reach the edge of its ability to comprehend. We are flesh and blood, not gods. Are there facets of the universe we can never know? Are there questions we can't ask? Our brains, which evolved to help us find food on the African plains, are not constructed to penetrate all the enigmas in the infinite mathematical cloak of our universe.

Deciphering the Mathematical Loom

> *Au fond de l'Inconnu pour trouver du nouveau. [Into the depths of the Unknown in quest of something new.]*
>
> —Charles Baudelaire, *Le Voyage*

In the fictional part of this chapter, the universe was likened to a mathematical carpet which is perceived by creatures looking at it through imperfect glasses. Computers may allow humanity to perfect those glasses. The beauty and importance of computers lie mainly in their usefulness as a tool for reasoning, creating, and discovering. Computers are one of our most important tools for reasoning beyond our own intuition.

We live in a civilization where numbers play a role in virtually all facets of human endeavor. Even in our daily lives we encounter multidigit zip-codes, social security numbers, credit card numbers, and phone numbers. In many ways the requirements for ordinary living are a great deal more complicated than ever before. Digits ... digits ... digits.... It all seems so dry sometimes. And yet, when one gazes at a page in a scientific journal and sees a set of complicated-looking equations in scientific texts, a sense of satisfaction is generated: *The human mind, when aided by numbers and symbols, is capable of expressing and understanding concepts of great complexity.* Ever since "visionary" mathematical and physical relations trickled like rain onto the rooftop of 20th-century humans, we have begun to realize that some descriptions of nature lie beyond our traditional, unaided ways of thinking.

The *expression* of complicated relations and equations is one magnificent step; *insight* gained from these relations is another. Today, computers with graphics can be used to produce representations of data from a number of perspectives and to characterize natural phenomena with increasing clarity and usefulness. "Mathematicians couldn't solve it until they could see it!" a caption in a popular scientific magazine recently exclaimed when describing work done on curved mathematical surfaces. In addition, cellular automata and fractals—classes of simple mathematical systems with exotic behavior—are beginning to show promise as models for a variety of physical processes. Today, in almost all branches of the scientific world, computer graphics are helping to provide insight and to reveal hidden relationships in complicated systems.

Today, computer calculations are now beginning to radically *change how scientists pursue and conceptualize problems*, and computer models open up entire new areas of exploration. In fact, of all the changes in scientific methodology, probably none is more important than the use of computers. The sheer amount of data generated by experiments is so large that comparisons and conclusions could not be made without computers. For example, massive DNA sequences have been uncovered—and only with the computer can hidden correlations be found within these bases in the genetic materials of organisms. Not unlike the search for extraterrestrial signals from space, scientists try to reconstruct messages and patterns in DNA strands, mathematical progressions, and a range of natural phenomena.

The remarkable panoply of computer applications is growing: Computers play a role in the design of other computers, in our understanding of molecular biology and evolution, and in processing images from outer space. The search for extraterrestrial intelligence employs the automatic detection of interstellar signals and requires sophisticated computers.

If the properties we assign to the natural world are partly expressions of the way we think and our capacity for understanding, then the introduction of new tools such as the computer will *change* those properties. The computer, like a microscope, expands the range of our senses. The world made visible by the computer seems limitless. In fact, in the next decade, almost all advances in science and art will rely partly on the computer and advanced technology. Moreover, humans will not be able to rely on any one single field of knowledge to make significant advances. Indeed, the computer of the 21st century will touch every aspect of our daily life, and enhance our creativity in all areas of artistic and intellectual expression. We'll all be peering into newly discovered regions of the mathematical loom—perhaps limited only by computers and our imagination.

CHAPTER 21

Eschaton Now

Nature does not know extinction; all it knows is transformation. Everything science has taught me, and continues to teach me, strengthens my belief in the continuity of our spiritual existence after death.

—Wernher von Braun

A screaming comes across the sky.
—Thomas Pynchon, *Gravity's Rainbow*

Your gaian spirit's like the Light—
contemptuous of clocks.
Our evanescent days and nights
mean nothing to the rocks.
But geophysics has its Tao,
and time will stop and stare
as you put flowers in your hair
in some eternal now.
  —Keith Allen Daniels, "Earth"

"Sir, where are we?"

Theano's hand is cold. "Where have you taken us?" she asks.

You flip on a light switch. "To the End of the World."

Mr. Plex spins around on his hindlimbs. "I–I can't believe it. We're back at the Temple of Apollo."

You nod. "August 15, 2126. We're going to observe Doomsday from here." You pause adjusting a view-screen on the wall. "I don't want to risk taking us any closer."

Mr. Plex paces back and forth. "This is madness, sir."

Theano takes a deep breath. "I'm not sure about this. I don't want to die today."

You put up your hands. "Don't worry. Seconds before our death, I'll whisk us away. We can live out the rest of our lives wherever and whenever we wish."

Theano suddenly stops and stares at a corner of the storeroom. "Tetraktys! What the hell is that?"

There is an ancient ribcage, a bit of mandible, and a splintered kithara.

You shrug. "Some poor soul must have gotten trapped in here."

Mr. Plex looks you directly in the eyes. "Where will the Swift-Tuttle comet hit?"

You press a button on a view-screen hanging on the wall. An image of a circular ring of rocks flickers onto the screen.

There is a haunted look in Theano's eyes. "Stonehenge?" she gasps.

"Yes, it's the precise impact site. In one hour, it will no longer exist."

Mr. Plex is unusually quiet. After a minute he says, "Sir, you're taking this all very calmly."

You shake your head. "Inside I'm a nervous wreck. But I have to see for myself how it will all end. Are our astrophysical predictions right? Is the mathematics correct? All through history, prophets of doom have predicted the end of the world, but none have been right. So far."

Another view-screen in the temple's backroom flickers to life. On it is a picture of Big Ben. The clock does not have hands. They were removed in 2015 when a Japanese corporation purchased the clock and installed a huge digital readout.

Big Ben reads 5:04:10.

Theano stares at the view-screen. "Maybe the Stonehenge builders somehow knew the comet would strike at Stonehenge?"

You turn to Theano. "Neat idea, but there's no evidence." You pause, adjusting a contrast knob on one of the view-screens. "I estimate that the comet will strike at 20 kilometers per second and vaporize Stonehenge upon impact."

You turn to Mr. Plex. "Could you send the electronic fly closer?"

He nods, wanders over to the other view-screen, and adjusts a dial.

Something in the sky is sparkling, very brightly. A new star. The blinding sphere begins to leave a short vapor trail in its wake.

"Sir!" Mr. Plex screams. "It looks too big to be the Swift-Tuttle comet! Could it be an asteroid?"

Big Ben reads 5:05:10.

The trail, no more than a smudge, rips in two directions.

Your eyes are glued to the screen. "I'm not sure," you say. "It does seem too large—"

"Shouldn't we be doing something? Sir? Run out into the towns? Warn the people in the cities? Call the President of the U.S.?"

You shake your head. "I'm sure his satellites and ground tracking systems know it's coming, but it's hard to stop something so large. They tried, but most of the world's nuclear arsenal was dismantled years ago. Budget cuts. World Peace, and all that.

The sirens are howling in the valleys near Salisbury.

Theano had been trembling, but now she is shaking violently. "It's glowing like fire!" she screams.

Your palms are sweating. "Friction with the air makes the surface incandescent."

The flying mass travels at a low angle near the ground, sputtering chunks as big as houses. A slender church steeple topples into the summer countryside.

A huge cylinder of superheated air is forced along by the projectile as it crashes to the ground. The air is pushed across the meadows in a flaming explosion that suddenly scorches every living thing for hundreds of miles in all directions.

Theano's voice is a thin croak that you barely recognize. "I can't believe it's happening," she says.

Another view-screen flickers to life.

A farmer in Salisbury, England stands upon his porch, looks up, and shields his eyes. The cosmic missile cuts into the atmosphere at a 30 degree angle above the horizon. At an altitude of about 6 km the object shatters in a rapid series of bursts and vaporizes, felling trees in a radial pattern over an area of 10,000 square kms and incinerating Stonehenge. It explodes with over 100 million times the force of the nuclear blast that devastated Hiroshima, Japan.

The Earth shakes like a vomiting god as the cosmic mass rips through a quarter of a mile of solid rock.

The crater created by the impact is an almost perfect circle, many miles in diameter. From the inside wall, the crater drops sharply to a depth of 1000 feet. Two Washington Monuments could be placed end-to-end in the middle of the Stonehenge crater.

Big Ben reads 5:06:66 and shatters.

The sky is a dusty red. Embers fall.

"Mon Dieu!" Mr. Plex screams in a high-sonic stiletto voice.

Theano is speechless.

Everything is still, but soon ashes are falling in large flakes, drifting with impossible slowness. Your electronic fly in England watches them drift down onto the fallen, leafless oaks and mistletoe, coating them with a thin, dull layer of greyness.

You wait a few hours. Time. What is time when there are no living animals in the forest? When ashes no longer fall on the fields? When the sky is filled with vague perpetual clouds?

But there is beauty in death: there are unusually colorful sunsets over western Europe. What is the old saying? Only that which can die is beautiful.

Theano is cadaverously pallid. "Why is the sky so–so strange?" she whispers.

A rope of nausea in your stomach knots tighter. "In the beginning, the night sky should be bright enough to cast shadows and allow a newspaper to be read. Ice-coated dust grains high in the sky create noctilucent (night-shining) clouds which illuminate the sky." You pause checking a readout at the bottom of one of the view-screens. "Disturbances in the Earth's magnetic field are reported 2700 kilometers southeast of the epicenter by ground observatories. These magnetic 'storms' are similar to ones produced by nuclear explosions in the atmosphere. A seismograph in St. Petersburg records tremors produced by the blast."

A siren is still screaming somewhere, like the cry of a desperate bird in a storm, a raven cawing in the deep pines, shooting up to the shattered crimson disk of the sun, toward the inverted sunset, and then plummeting to the ground like a shooting star.

毅 毅 毅 毅

Sudden violence upon the air and no trace afterwards ... a Word, spoken with no warning into your ear, and then silence forever. Beyond its invisibility, beyond hammerfall and doomcrack, here is its real horror ... beaten like Death's drum, still humming.... God, where'd that come from?

—Thomas Pynchon, *Gravity's Rainbow*

"Sir, send out the flies. Show us what's happening!"

You nod, and press a button on the view-screen.

A large dust cloud high in the atmosphere catches the sun's rays and reflects them downward, even after the sun has sunk below the horizon. There are brilliant red sunsets throughout the world, like blood flowing from a dying god.

The charred remains of British royalty decorate the hillside.

In the grey waters of the Mediterranean, an Italian head of state wields a long knife and chases a nude woman through swamps the color of tabasco.

In the peaceful suburbs of Atlantic City, the ex-governor of New Jersey is living off road kill, and now calmly barbecuing his neighbor's cat.

Unsanitary conditions are everywhere. Acute pigbel, a bacterial disease, kills thousands as massive clostridial growths balloon intestines and destroy voluntary muscles in a process called rhabdomyolysis.

Civilization is fizzing away like bubbles in an open champagne bottle.

"Incredible," Mr. Plex moans.

Theano points to the view-screen. "What the hell?" Her voice trembles. Something is stalking through the English countryside.

Mr. Plex screams. "It—it's the transfinites.

Zeus, Aphrodite, Poseidon, Apollo, and the Seven Sleepers pour out through a crack in the Earth. They walk through the smoke, gathering up the dead. Their heavy footsteps rumble like a digestive system undergoing extreme peristalsis.

"Sir, what are these transfinites really?"

"We've tried to figure that out for decades. All religions seem to acknowledge the presence of discarnate entities. Maybe the Catholic Church thought of them as demons. But they're really time travelers without bodies, aliens."

"But in the cave. Wasn't that what they really looked like?"

"Could have been just another ruse."

For a moment, Mr. Plex's abdomen appears to turn crystalline and crumble, the sparkling broken shards of glass reflecting a myriad of colors which tinkle and fade in the still air. But it's only a vision. Not reality.

Mr. Plex's forelimbs begin to tremble. "Why do they animate statues of Greek gods?"

You shrug. "Maybe because that's where we first visited. Ancient Greece."

The globule on your statue of Dionysus' knee is glowing. You tear it from the statue, throw it on the ground, and crush it with your shoe. "I'm worried the transfinites could use it to trace our location."

As Theano gazes at the destruction on the view screen, a tear comes to her eye.

You walk over to her, thinking that you might as well tell her the truth. "When the comet hit, it excavated a lot of ground. If just 1 percent of the mass reaches the upper atmosphere, the entire Earth will soon be covered with a layer of fine dust several centimeters thick." You pause. "It will soon be totally dark, even in the day. Part of the atmosphere was blown away. And a few huge fragments which landed in the ocean are causing huge tsunamis. The oceans around England are nearly boiling."

"Sir, for how long can humans survive?"

"Chemical reactions catalyzed by the comet's journey through our atmosphere will generate acid rain." You pause pointing to the screen. "Fires are springing up in the forests.

At first, there will be crop damage, later worldwide starvation. Economics systems, government, and civilization will be destroyed."

Theano's eyes appear to sink into her skull and lose much of their luster. "But could humans survive?"

"Initially, yes." You look down. "But eventually most higher lifeforms will die. The earthquakes, tidal waves, volcanic action...." You look back at Mr. Plex. "I told you that some of the Earth's atmosphere was knocked right off into space. There will be enormous amounts of debris and carbon dioxide in whatever's left. The sea will be polluted."

You kick at the ancient rib cage in the corner of the storage closet of the Temple of Apollo. "Small pockets of humans will hang on for a while, but they'll slowly die out as social systems break down. Climactic changes kill crops. The cows and horses die. Plagues spread. The last to go will be the cockroaches." You pause. "Still there is hope. Some of the settlers on Mars and the Moon should be able to survive for a long time if they manage their resources properly."

Theano gazes at the view-screen. "Who the hell is that?"

On the screen is a man. Apparently drunk. Dancing to Michael Jackson's *Beat It*.

"I've sent an electronic fly three weeks into the future and through an air vent in an underground bunker near Washington, D.C." You pause. "That man is the President of the United States of America. He knows his food supplies won't last indefinitely."

The President is smoking a strange cigarette as he reaches for a family-size bottle of valium. He has evidently become such a habitual drug user that his complexion has acquired a waxy appearance.

Mr. Plex is looking at the floor. His movements are slow. "Sir, I don't want to stay around on Earth any longer."

When you listen very carefully, you hear the faint sound of ocean waves. Waves pounding against large craggy outcroppings of rocks.

There are tears in Theano's eyes. "Let's get out of here."

The air is becoming more difficult to breathe.

"Sir, we don't have much time."

You look at Theano and notice red blotches on her skin. They seem to bloom with the regularity of a dripping faucet. "What's happening to your skin?"

"Not sure," Theano says. She carefully examines her arms. "Maybe an allergic reaction. I do have nasal allergies, but nothing like this has happened before. Probably no big deal."

This is no time for a medical emergency. The floor is trembling and you hear the churning of water somewhere in the distance. You look down. "It's spreading to your legs. It *is* a big deal."

"I–I can't seem to concentrate," Theano says as her eyes get red. She sucks her mouth into a rosette. "My eyes feel puffy. Maybe I'm just being a hypochondriac."

"No way. This is serious."

Theano's lips are beginning to resemble pink swollen pickles. She has trouble speaking or swallowing. "Mr. Plex, hold on to her. I think I have some medicine in my pack."

Theano's eyes are swollen shut.

Mr. Plex cloaks her in his arms. "Sir, what is it?"

"Anaphylaxis."

劋

"What's that?"

"It's the Schwarzenegger of allergic reactions."

"Sir, how dangerous?"

"Her immune system is attacking something that got in her. Maybe it's something stirred up by the comet."

Theano grabs her stomach. "I feel a terrible cramp," she says.

The floor begins to buckle slightly as if some gigantic mass were applying pressure from beneath. "Sir, what do we do?" Mr. Plex yells.

You riffle through your pack and pull out a tiny ampule of hydrocortisone and adrenaline. "My God, I didn't think this was here. It's old, but it should quiet the spasm of her airways and stomach."

A minute passes. You rest Theano on the floor. "Can you breathe better now?"

"Yes," she says. Her voice is weak, but free of fear and pain. In another minute she is almost back to normal.

The air has become very damp, and from outside you hear the swirling of waves. Perhaps a tidal wave. "Let's get out of here!" Mr. Plex screams.

You nod. "Mr. Plex, where will you go?"

"Back in time to Mrs. Plex. She's waiting for me on our intergalactic museum floating in outer space."

You nod. "You're in charge until we get back."

Mr Plex shambles closer to you. "And you, Sir?"

"Yes?"

"Where will you and Theano go?"

You smile at Theano and turn back to Mr. Plex. "We'll send you a postcard."

劋 劋 劋 劋

It may have been a human figure, dreaming of an early evening in each great capital luminous enough to tell him he will never die, coming outside to wish on the first star. But it was not a star, it was falling, a bright angel of death.

—Thomas Pynchon, *Gravity's Rainbow*

A monkey on the African savannah is crying.

A golden tamarin whimpers.

A flamingo shrieks.

Large mammals in the wild are the first to go. And then birds, followed swiftly by reptiles. A kind of de-evolution is taking place. A few tiny arthropods live out their meager lives amidst the gray mists and growths. Least affected are insects.

A few thousand thermophilic worms living along hot deep sea vents survive for awhile but even they do not like acid oceans. At least the primitive archaea microbes don't seem to give a damn. These hot-spring creatures merrily estivate in Obsidean Pool—a bubbling dark caldron 9 feet by 27 feet in size. Their enzymes withstand heat, acids, and salt.

The marshy sea, mother of life, now stands choked with fungoid growths and goo. A solitary sea creature slithers along grey surfaces of rock, and with a last spasm of terror, fades into fetid chasms of empty air.

Long after *The Urantia Book*, *Ars Magna* of Ramon Lull, and Mandelbrot's *Fractal Geometry of Nature* have turned to dust, will an intelligent civilization of worms, arachnids, or reptiles evolve? Will Pythagoras' legacy rise like a phoenix from the ashes given sufficient time and intelligent creatures? Perhaps a wormlike archeologist will discover a golden tetraktys or petrified kithara and wonder about its meaning, holding them up before an incredulous convention of nematodes or offering it on the altar of the worm gods.

Einstein, an intelligent iguana.

And Pythagoras, a worm.

THE SCIENCE BEHIND THE SCIENCE FICTION

Considering the significant number of asteroid bodies potentially on collision courses with Earth, a repetition of the Tunguska event is virtually certain.
—Roy Gallant, *Sky & Telescope*, 1994

We orbit the Sun within a sparse swarm of asteroids and comets, some of which will ultimately and inescapably collide with our planet.
—David Morrison, *Sky & Telescope*, 1990

The hand of God emerges from a cloud, the edges of the figure here and there eroded by 200 years of season's fire and ice chisels at work.
—Thomas Pynchon, *Gravity's Rainbow*

Tunguska and the Wrath of Ogdy

Projectiles from space have slammed into the Earth in the past. For example, we know that on the morning of June 30, 1908, a fireball plunged down into the Siberian sky over Tunguska, and exploded with 2000 times the force of the nuclear blast that devastated Hiroshima, Japan. The projectile weighed between 100,000 and 1 million metric tons and, at an altitude of about 6 km, shattered in a rapid series of bursts and vaporized, felling trees in a radial pattern over an area greater than 2000 square km. It burned a central area half that size (see p. 214).

Your observations with Mr. Plex and Theano regarding sunsets, magnetic storms, and night-shining clouds are all based on scientific observations made following the 1908 Tunguska explosion. After the impact, colorful sunsets were seen in Russia, Scandinavia, and even in Western Europe. A seismograph in St. Petersburg recorded tremors produced by the blast. The night sky was bright enough to cast shadows and allow a newspaper to be read. Ice-coated dust grains high in the sky created noctilucent (night-shining) clouds that illuminated the sky.

Disturbances in the Earth's magnetic field were reported 2700 km southeast of the Tunguska epicenter by ground observatories. (These magnetic "storms" are similar to ones produced by nuclear explosions in the atmosphere.)

The Tunguska impact left thousands of burned reindeer bodies and felled 80 million trees. The projectile may have been a comet or asteroid several miles in diameter weighing a million tons. Today, we can only speculate about the exact nature of what caused the

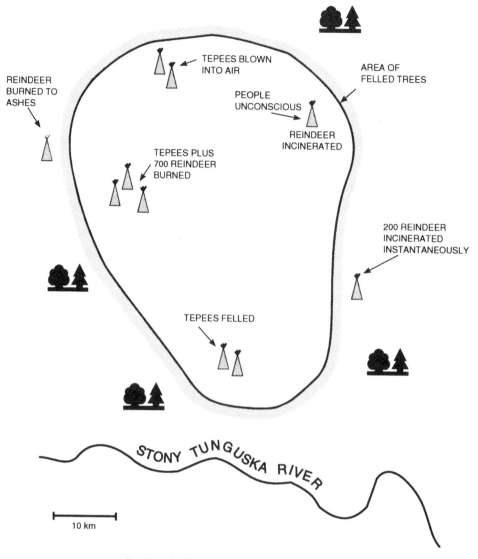

Surviving the Tunguska blast (After Gallant, 1994).

Tunguska event. No remnant of comet or meteorite has ever been found. The projectile's impact was at exactly the same latitude as the city of St. Petersburg. If the comet plunged to Earth just two hours earlier than it did, it would have hit the city and killed every inhabitant. If a St. Petersburg collision had occurred during the cold war, would it have triggered a nuclear war before the superpowers had a chance to determine the Apocalypse had come from outer space?[1]

As is often true of disasters, many of the local people attributed a religious significance to the Tunguska impact. For example, following the explosion, local residents considered the region off-limits, a result of the wrath of Ogdy the fire god who had punished them for their disobedience. Travel through the region was subsequently forbidden.

The Earth exists within a cosmic shooting gallery: impact-induced catastrophes have been part of its natural history for billions of years. As examples of some close calls, consider that in March 1989 a large asteroid passed within 100,000 km of Earth. Had it struck Earth, the impact would have been the equivalent of more than a million tons of exploding TNT and created a crater 7 km across. This asteroid is just one of thousands in Earth-approaching orbits. Amazingly, fewer than 80 Earth-approaching asteroids have been discovered to date, although sampling statistics imply that many more fly past Earth undetected. (Table 21.1)

Cosmic impacts occur throughout our Solar System. David Morrison of NASA-Ames Research suggests the best evidence for collision-induced catastrophes comes from the Moon's battered surface. (If the Earth's crust were not continually being eroded and pushed by underground geological activity, the face of our planet might more closely resemble the pitted Moon.) In fact, using the Moon to estimate the frequency and magnitude of impacts on Earth, the Earth endures an impact comparable to those that formed the five largest lunar mare craters once every 10 million years (see p. 216).

We can estimate the size *distribution* of impacts by counting the number of lunar craters in different sizes ranges. It turns out that the number of craters of a given size is roughly inversely proportion to the diameter squared (see p. 217). For example, craters 25 to 50 km across are about five times more common than those 50 to 100 km across. Also, the time between impact events of a given size varies directly as the diameter squared. A 1 km object strikes the earth about 100 times more often than those 10 km across. Even though larger impacts are more rare than small ones, such events can't

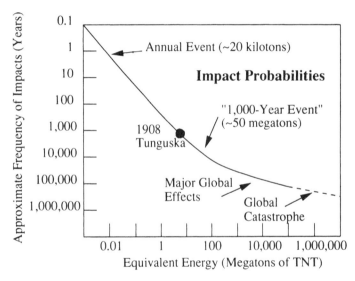

Impact probability energies, and frequency of occurrence. This plot indicates the chances our planet will be a hit by a member of the current population of Earth-crossing asteroids and comets. Notice that larger impacts are less common than smaller ones. Every year the odds are one in some hundred thousand of having a civilization-threatening impact. (After Morrison, 1990).

Table 21.1 Some of the larger Earth-crossing asteroids. Notice that most were discovered in the past 40 years.

Number	Name	Yr discovered	Radius (km)	Period (yrs)
1685	Toro	1948	2.4	1.60
2212	Hephaistos	1978	4.4	3.18
4015	1979VA	1979	1.6	4.29
1864	Daedalus	1971	1.7	1.77

The lunar crater Copernicus and its surroundings in afternoon illumination. Examples of every type of lunar surface feature are shown on this plate taken with a 36-inch refracting telescope. (From Cherrington).

be avoided. Just look at the Moon. The cataclysm at the end of the Cretaceous destroyed most life on Earth. Most families of marine animals were wiped out, and there was widespread extinction of land animals and plants.

The figures on pp. 217 and 218 are my computer graphical simulations of a roughly $1/r^2$ distribution of craters. See the Appendix program code for details.[2]

What would happen if a comet were to hit the Earth? For one thing, the comet would be travelling at tens of kilometers per second and would be vaporized upon impact. Even if just one percent of the excavated Earth reaches the upper atmosphere, the entire Earth would soon be covered by a fine layer of dust several centimeters thick. The Earth would become a dark world. In addition, a large impact could actually blow away part of our atmosphere. If the strike occurs in the ocean, huge tsunamis would be produced. Acid rain would result from chemical reactions catalyzed by the comet passing through the atmosphere.

Strangely, there is a good side to cometary impacts. Without impact-generated diversification of animals, there might be less opportunity for development of life as new species evolve from survivors. If the dinosaurs did not die out, humans would not exist.

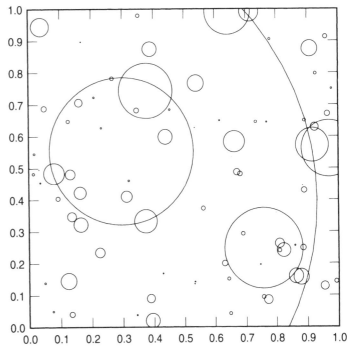

Computer simulation of crater size distribution.

A Note on Swift-Tuttle

Because of the impossibility of accounting for the microdynamics of the comet's motion, we don't know how any error in the predictions will propagate.
— Brian Marsden, *Sky & Telescope*, 1993

I'm intrigued that the possibility of a strike by Swift Tuttle—admittedly remote—should arise at a time when there is unparalleled international interest in defending ourselves against just such a threat.
— Brian Marsden, *Sky & Telescope*, 1993

The Swift-Tuttle comet is discussed in Chapter 11. As author Brian Marsden notes, depending on the size and density of the comet's nucleus, which we don't know, such an encounter could precipitate the end of human civilization through nuclear winter scenarios. On the remote possibility that the comet plummets to earth sometime around August 14, 2126, jets of dust spreading tens of thousands of miles will engulf the Earth and the Moon.

THE LOOM
OF GOD

And the stars of heaven fell unto the earth, even as a fig tree casteth her untimely figs, when she is shaken of a mighty wind.

—Revelation, 6:13

As mentioned in other sections, there are many asteroids in orbits which intersect the orbit of the Earth. From where do these harbingers of doom come? How large can they be?

Within the asteroid belt between Mars and Jupiter fly asteroids like Ceres and Pallas, 480 and 300 miles in diameter. Some scientists believe that all the fragments in the asteroid belt are the remains of a planet that either broke up or somehow failed to be born. A few large asteroids have escaped from the belt, and all of these escapees have been given masculine names, usually after Greek gods and heroes. For example, in 1968, asteroid Icarus set off a Doomsday scare when it came close to the Earth. If it hit, it would have been the equivalent of 7000 100-megaton thermonuclear bombs. On October 39, 1937, asteroid Hermes (see p. 219) nearly collided with the Earth, coming within a mere 485,000 miles—twice as far away as the moon.

About 130 impact craters have been discovered on Earth. For example, one large crater in Arizona is 4000 feet in diameter and 600 feet deep. Scientists believe the crater was created 50,000 years ago by an asteroid about 80 feet in diameter weighing about 10,000 tons.

I recently used computer programs to simulate the effect of different sized asteroids colliding with the Earth. The following results come from *Discover Space* software created by Broderbund Software (1993). The program allows

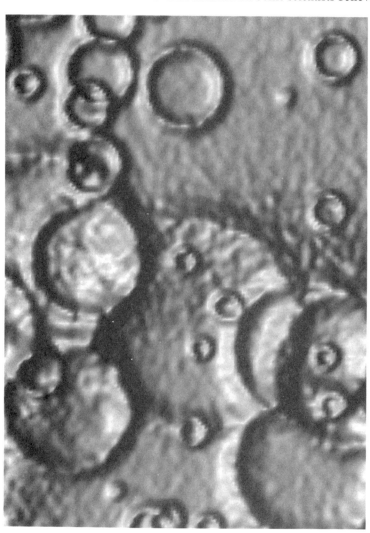

Three-dimensional computer graphics simulation of lunar craters.

users to enter values for asteroid diameters and impact velocities, and then view the resultant carnage when the asteroid slams into the Earth. I have chosen a range of numbers to give you an indication of a variety of outcomes. In the following table, we consider 6 asteroids, *A* through *F*, having different diameters and impact velocities.

	Asteroid Diameter (km)	Impact Velocity (km/sec)	Crater Size	Explosive Yield
A	0.1	2	.5 km	43 Hiroshima bombs
B	1	20	21 km	4.4 million Hiroshima bombs
C	1	40	31 km	17.5 million Hiroshima bombs
D	10	30	201 km	9.9 billion Hiroshima bombs
E	20	20	292 km	35 billion Hiroshima bombs
F	20	40	439 km	140 billion Hiroshima bombs

Asteroid A. The impact of a small, slow-moving asteroid, as in scenario A, would create damage equivalent to the detonation of a thermonuclear warhead. (The asteroid's mass of 1.8 million metric tons, equal to a coal train 173 miles in length, would have an explosive yield of 876 kilotons of TNT.) If such an asteroid were to plunge into the headquarters of the Internal Revenue Service in Washington, D.C., none of the buildings from the White House to the Capitol Building would be left standing. Such an impact would flatten everything for miles and kill thousands.

Asteroid B. An asteroid of the size and velocity in scenario B is expected to hit the Earth once every thousand years, throwing rocks, dust, and water vapor high into our atmosphere. (The asteroid's mass of 1.8 billion metric tons, equal to a coal train 173,000 miles in length, would have an explosive yield of 87 thousand megatons of TNT.) If such an asteroid plunged into San Francisco, the entire Bay Area would be destroyed and the San Andreas fault could be triggered by the shock. All humans within hundreds of miles would be killed by the overpressure blast of the shock wave and by radiation produced by a plasma sheet during atmospheric entry.

Asteroid C. Like Asteroid B, Asteroid C is expected to hit the Earth roughly once every thousand years and would throw rocks, dust, and water vapor high into our atmosphere. (The asteroid's mass of 1.8 billion metric tons, equal to a coal train 173,000 miles in length, would have an explosive yield of 350,000 megatons of TNT.) See Asteroid B description for lethal effects.

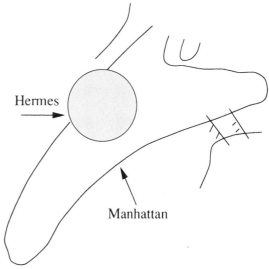

Schematic illustration showing comparative size of the meteoroid Hermes, and Manhattan Island.

Asteroid D. The impact of an asteroid as in scenario D would produce a global catastrophe beyond our wildest imaginations. (The asteroid's mass of 1.8 trillion metric tons, equal to a coal train 173 million miles in length, would have an explosive yield of 197 million megatons of TNT.) If such an asteroid plunged into Earth, our climate would change due to the large amount of water vapor, dust, and nitric acid blasted into the air. Earthquakes would be set off around the world, and ejected rocks and molten lava would rain down upon the cities. An impact of this type may have been responsible for killing the dinosaurs 65 million years ago. Another mass extinction occurred 250 million years ago and ended the Permian era. This extinction was even more devastating than the impact that wiped out the dinosaurs and may have been impact triggered. Impacts have come uncomfortably close to sterilizing our planet of larger life forms on a number of occasions in our past.

Asteroid E. This asteroid would have an effect similar to Asteroid D. (The asteroid's mass of 14.7 trillion metric tons, equal to a coal train 1.4 billion miles in length, would have an explosive yield of 701 billion megatons of TNT.)

Asteroid F. This asteroid would also have an effect similar to Asteroid D. (The asteroid's mass of 14.7 trillion metric tons, equal to a coal train 1.4 billion miles in length, would have an explosive yield of 2.8 billion megatons of TNT.) Shown below is the crater size produced by this asteroid superimposed on the state of New Mexico.

Impact velocities higher than those listed in the table are likely for many asteroids. Note also that if an object with a 100-mile diameter slammed into the Earth, it would dig a crater the size of the Pacific Ocean. Could such a collision have created the Pacific Ocean basin early in the Earth's history?

New Mexico

Asteroid F

The crater size produced by Asteroid F superimposed on the state of New Mexico.

COMETS AND GODS

God protect us from the Comet and the Fury of the Norsemen.
—Medieval prayer

When beggars die, there are not comets seen; The Heavens themselves blaze forth the death of princes.
—Shakespeare, *Julius Caesar*

In 1910 the appearance of Halley's comet (see p. 221) caused quite a stir, and some group predicted it would destroy the Earth. (The Earth, in fact, did pass through the gaseous tail of the comet.) One Englishman with an entrepreneurial spirit sold comet pills—a mixture of aspirin and sugar—to ward off the effects of the poisonous gases. Pseudo-religious cults briefly developed as local groups banded together to commiserate about the End of the World.

Throughout history, comets and God have always gone hand in hand. To Christians and Moslems of the Middle Ages, comets were usually considered examples of God's wrath. Consider the following listing of comets I found in a medieval Christian document that linked comets to the Devil:

Anno 1531, 1532 and 1533 [comets] were seen, and at that time Satan hatched heretics.

In A.D. 69, Jewish rebels in the city of Jerusalem looked up into the sky and saw a comet. At the time, the rebels were fighting a difficult battle against the Romans who were trying to retake the city. Unfortunately for the Jews, their prophets urged them to climb the roof of the Temple to see the miraculous sign of their coming deliverance by God. The Jewish prophets were in error: The city was captured by the Romans the following year, and the holy Jewish temple was burned. Would the battle have had a different outcome if the Jews had not attributed a religious significance to the comet?

Drawing of Halley's Comet in 1910.

COMET ANATOMY

If we were to send Mr. Plex with his strong diamond body out into space to explore the interior of a comet, what would he find? For one thing, the central body of a comet is like a dirty iceberg containing frozen methane, ammonia, and perhaps water. Embedded in the gasses are small, solid, stonelike particles. Here are some other facts:

- Comets melt away through time, fading after a few thousand years. When a comet's tail passes by Earth, we encounter some of the small bits of solid material causing a meteorite shower.[3]

- We have nothing to fear from a comet's tail. There is more material in a cubic inch of air above a Manhattan street than there is in 2000 cubic miles of a comet's tail.

- The main body of an average comet is 1.2 miles in diameter. Very large comet bodies can be many hundreds of miles in diameter, and if such a comet hits the Earth the results would be catastrophic. Lucky for us, the probability of such a comet hitting is low, once every 100 million years.

- Like our own planet, comets orbit around the sun. Unlike the Earth, a comet's orbit is usually less regular and often travels far beyond the orbit of Pluto. Comets generally have quite long orbital periods. For example, Comet Encke has the shortest orbital period of any comet known: 3.3 years.

The Black Stone: A "Bomb from Heaven"

The bay-trees in our country are all wither'd
And meteors fright the fixed stars of heaven;
The pale-faced moon looks bloody on the earth
And lean-look'd prophets whisper fearful change, ...
These signs forerun the death or fall of kings.
—Shakespeare, *Richard II*

It would be possible to test hypotheses of origin, such as, for example, the idea that some 5 million years ago, about the time of the origin of the hominids, the Kaaba was chipped off an asteroid named 22 Kalliope ... and accidentally encountered the Arabian Peninsula 2,500 years ago.

—Carl Sagan, *Broca's Brain*

Due to their sheer numbers, meteorites are much more dangerous to Earth than comets. Harvard Observatory estimates that 100 billion meteroids hit our atmosphere every day.

Meteorites were once considered "bombs" from heaven, and were often worshipped. Do you think the following biblical passage refers to them?

Men of Ephesus, what man is there who does not know that the city of Ephesians is temple keeper of the great Artemis, and of the sacred stone that fell from the sky? (Acts 19:35)

Meteorites have also been discovered in Aztec temples, and in North America they have been buried with the corpses of Indians. The oldest example of meteorite worship comes from one of the seven wonders of the ancient world: the Temple of Diana at Ephesus in Asia Minor. Residing in its exquisite Greek chambers was a black rock, probably metallic, an object from outer space.

In 1492, a 300-pound meteorite fell near the army of Emperor Maximilian who ordered that the stone be carried to his castle. His councilors declared that the meteorite was sign of God's favor, and the Emperor had the stone hung in the parish church in Ensisheim.

Some scholars suggest that the Black Stone of the Kaaba in Mecca—the most sacred object in Islam—is a meteorite. In fact, stone worship was common in pre-Islamic Saudi

Arabia. Arab merchants controlling lucrative trade routes also managed lucrative religious rituals that centered around the Kaaba and its sacred Black Stone.

The Kaaba today is a rectangular stone edifice (40 × 35 × 50 feet). The Black Stone is embedded in its southeast corner, five feet from the ground. The stone is actually made of a dark red material, oval in shape, some seven inches in diameter. In ancient days, many of its worshipers kissed the stone to show the depth of their feeling.[4]

FOUR RECIPES FOR A DOOMSDAY MACHINE

We have so far been discussing Doomsday brought about by natural causes. But what about Doomsday Machines? The term *Doomsday Machine* refers to the class of hypothetical weapons specially designed to destroy all large lifeforms including humans. Could such a weapon be produced? Sadly the answer may be yes. Here are four recipes based on the speculations of Daniel Cohen, author of *Waiting for the Apocalypse*:

1 *Cobalt Bomb Cluster*. The easiest Doomsday Machine to construct is the cobalt bomb cluster. Each cobalt bomb is an ordinary atomic bomb encased in a jacket of cobalt. When a cobalt bomb explodes, it spreads a huge amount of radiation. If enough of these bombs were exploded, life on Earth would perish.

2 *Wobble Bombs*. In this recipe for Doomsday, large hydrogen bombs are placed at strategic locations on Earth and exploded simultaneously. As a result, the Earth may wobble on its axis. If placed at major fault lines, the bombs could trigger a worldwide series of killer earthquakes.

3 *Asteroid Bomb*. This Doomsday device was actually suggested by Dandrige Cole, a speculator for the General Electric Corporation. He postulated it is possible to capture one of the larger asteroids and send it crashing to Earth by exploding nuclear bombs at specific locations on the surface of the asteroid.

4 *Botulis Bombs*. Biological Doomsday Machines refer to biological weapons utilizing bacteria, viruses, or various biological toxins. For example, a few pounds of poison produced by the botulis bacteria is sufficient to kill all human life.

Other Ways to Die

The universe faces a future extinction of endless cold or intolerable heat. The more the universe seems comprehensible, the more it also seems pointless.
—Steven Weinberg, *The First Three Minutes*

Some say the world will end in fire. Some say in ice …
—Robert Frost

Many believe that the Earth is like an inmate waiting on Death Row. Even if we do not die by a comet or asteroid impact, we know the Earth's days are numbered. The Earth's rotation is slowing down. Far in the future, day lengths will be equivalent to 50 of our present days. The Moon will hang in the same place in the sky, and the lunar tides will stop.

In five billion years, the fuel in our Sun will be exhausted. The Sun will begin to die and expand, becoming a red giant. At some point, our oceans will boil away. No one on

Earth will be alive to see a red glow filling most of the sky. As Freeman Dyson once said, "No matter how deep we burrow into the Earth ... we can only postpone by a few million years our miserable end."

Where will humans be, five billion years from now, at the End of the World?[5]

Even if we could somehow withstand the incredible heat of the Sun, we would not survive. In about 7 billion years, the Sun's outer "atmosphere" will engulf the Earth. Due to atmospheric friction, the Earth will spiral into the sun and incinerate.

Here is our future:

Sun expands to engulf the Earth.	7 billion years
Stars cease to form. All large stars have become neutron stars or black holes.	1 trillion years
Longest-lived stars use up all fuel.	100 trillion years

And when he opened the seventh seal, there was silence in heaven about the space of half an hour.

—Book of Revelations (8:1)

Is There Hope?

If this ending seems too dismal to you, perhaps we should ask if there is hope for humanity when the Sun expands to engulf the Earth in seven billion years. To give an answer, first consider that around four billion years ago, living creatures were nothing more than biochemical machines capable of self-reproduction. In a mere fraction of this time, humans evolved from creatures like Australopithecines. Today, humans have wandered the Moon and have studied ideas ranging from general relativity to quantum cosmology. Once space travel begins in earnest our descendents will leave the confinement of Earth. Because the ultimate fate of the universe involves great cold or great heat, it is likely that *Homo sapiens* will become extinct. However, our civilization and our values may not be doomed. Who knows into what beings we will evolve? Who knows what intelligent machines we will create that will be our ultimate heirs? These creatures might survive virtually forever, and our ideas, hopes, and dreams carried with them. There is a strangeness to the loom of our universe which may encompass time travel, higher dimensions, quantum superspace, and parallel universes—worlds that resemble our own and perhaps even occupy the same space as our own in some ghostly manner. Our heirs, whatever or whoever they may be, will explore these new regions of the loom. They will explore space and time. They will seek their salvation in the stars.

I conclude on an upbeat note from theoretical physicist Freeman J. Dyson

> Goedel proved that the world of pure mathematics is inexhaustible; no finite set of axioms and rules of inference can ever encompass the whole of mathematics; given any finite set of axioms, we can find meaningful mathematical questions which the axioms leave unanswered. I hope that an analogous situation exists in the physical world. If my view of the future is correct, it means that the world of physics and astronomy is also inexhaustible; no matter how far we go into the future, there will always be new things happening, new information coming in, new worlds to explore, a constantly expanding domain of life, consciousness, and memory.

CHAPTER 22

Epilogue

Sometimes it's a form of love just to talk to somebody that you have nothing in common with and still be fascinated by their presence.
—David Byrne

When an electron vibrates, the universe shakes.
—British physicist Sir James Jeans

While the great Loom of God works in darkness above ...
—Thomas Pynchon, *Gravity's Rainbow*

The Agean Sea is ebbing onto an island beach. Overhead, a solitary sea gull cries and is soon joined by several others.

Theano emerges from a cabin about twenty feet from the surf. Her feet are bare. Her skin is no longer pale. Her hair flows like silk in the breeze. "Do you have to keep that?" she asks.

"Dionysus?"

"What else?"

"Just a memento—"

"We don't need it."

You shrug. "You're right."

You get up, walk to the edge of the surf, and heave the mutant Dionysus into the infinite ocean. The lovely pounding of the surf is the only sound. You smile as a pleasant ozonic smell of low tide fills your nostrils.

The churning surf: swirls upon swirls upon swirls. There is a fractal heaven, you think. This is your new home.

From the corner of your eye, you see Theano with her long hair and sparkling eyes. She has climbed into a small Mytilenian boat with the Greek letters Μυτιληναιοσ on its side. As she turns it toward you, you feel a giddy excitement. It must be ecstasy to be riding on this cool, vast ocean in such a small craft. A slender spiral wave caresses her smooth skin, and she laughs.

The spiral waves split light into a million colored sparkles as Theano guides her craft toward you and waves her hand. You think her smile is radiant. You wave back and then look above her head at the fractal clouds. "Freedom," you whisper as fiery orange light reflects from a logarithmic spiral seashell, challenging the azure of an endless fractal sky.

The craft drifts closer and you climb in. You want to say something to Theano, something profound, something that adequately expresses your emotions. Your heart beats faster, and suddenly you turn to her, and gently hold her hand.

"I know," she whispers, giving your hand a squeeze.

You think she looks like a goddess in the glorious explosion of rosy light. She is wearing the quipu around her neck. Her hand feels warm.

"Let's go," she says, tenderly guiding you into her craft. Together you gaze into the shining and limitless heaven. The craft turns and points to a symmetrical collection of intertwined aquamarine spirals.

You are moving toward open sea.

"I have something for you," she says reaching under the seat. She hands you a crystal glass of Kylonian brandy.

"Cheers," she says with a smile as she clinks her glass into yours.

You both gaze up into the sky and spot ten seagulls, wings outstretched and motionless in tetraktys formation, flying toward the amber disk of the sun.

The sunset is like a flock of seagulls on fire.

> *The mathematical spirit is a primordial human property*
> *that reveals itself whenever human beings live*
> *or material vestiges of former life exist.*
> —Willi Hartner

POSTSCRIPT **1**

Goedel's Mathematical Proof of the Existence of God

Were theologians to succeed in their attempt to strictly separate science and religion, they would kill religion. Theology simply must become a branch of physics if it is to survive. That even theologians are slowly becoming effective atheists has been documented ...

—Frank Tipler, *The Physics of Immortality*

God created the natural numbers, and all the rest is the work of man.
—Leopold Kronecker (1823–1891)

In Chapter 20, I alluded to the famous mathematician Kurt Goedel and the fact that he spent the last few years of his life working on a mathematical proof of God's existence. In this Postscript, I quote colleagues from around the world who have responded to my questions on this subject, and I thank them for permission to reproduce excerpts from their comments. Some of my questions were sent through electronic mail or posted to electronic bulletin boards. Three common sources for such information exchange were "sci.math," "alt.atheism," and "soc.religion.christianity"—electronic bulletin boards (or newsgroups) that are part of a large, worldwide network of interconnected computers called Usenet. (The computers exchange news articles with each other on a voluntary basis.)[1]

Without further ado, I present Kurt Goedel's proof of God's existence:

GOEDEL'S MATHEMATICAL PROOF OF GOD'S EXISTENCE

Axiom 1. (Dichotomy) A property is positive if and only if its negation is negative.
Axiom 2. (Closure) A property is positive if it necessarily contains a positive property.
Theorem 1. A positive property is logically consistent (i.e., possibly it has some instance).
Definition. Something is God-like if and only if it possesses all positive properties.

Axiom 3. Being God-like is a positive property.

Axiom 4. Being a positive property is (logical, hence) necessary.

Definition. A property P is the essence of x if and only if x has P and is necessarily minimal.

Theorem 2. If x is God-like, then being God-like is the essence of x.

Definition. $NE(x)$: x necessarily exists if it has an essential property.

Axiom 5. Being NE is God-like.

Theorem 3. Necessarily there is some x such that x is God-like.

Note: I obtained this proof from: Wang, Hao (1987) *Reflections on Kurt Goedel*. MIT Press: Cambridge, Mass. (page 195).

INTERNET INTERPRETATION OF GOEDEL'S PROOF OF GOD

Christians grossly underestimate the pleasures that will be available in Heaven. They greatly underestimate what a being with literally infinite power can do.
—Frank Tipler, *The Physics of Immortality*

The appeal of mathematics for artists is precisely the universality of mathematics— universal in the sense of culturally invariant. This is not mathematics as a tool to prove God but rather mathematics as evidence of some universal fabric for existence.
—Anonymous Internet Traveler

You can't disbelieve something if you don't even know what it is you're not believing in. Likewise, you can't believe in something if you don't even know what it is you're believing in. This makes everyone an agnostic until or unless a definition is agreed upon.
—Mark Hopkins

I could not find significant information in the philosophical or mathematical literature regarding the interpretation of Goedel's proof of God. Author Hao Wang expressed no opinion or help to the reader in interpreting the proof. In fact, he states enigmatically, "I am only including [the proof] for some measure of completeness, and shall, out of ignorance, make no comments on it." I therefore consulted various mathematicians and philosophers on the Internet for their opinions. If you don't understand all of their comments, don't worry. At the least, they will give you the flavor of how current researchers in philosophy and mathematics express their thoughts on complex theomatic areas. I have not eliminated repetitious information so that you can get a feel for the overlap of current, common thinking among respondents.

Ivan Ordonez Reinoso, a doctoral student in Computers and Information Sciences at the Ohio State University, comments:

> The proof has an obvious problem in that it only shows the existence of a mathematical object with the property "God-like" as defined in the proof. There is no relation to the real world; it is like proving that there is no last cardinal number. Furthermore, the proof does not exclude the possibility of the existence of an infinite number of disjoint objects all with the property "God-like."

Mike Miller from Western Illinois University comments:

> Here's the easiest place at which the proof may be ripped apart. Nowhere is it shown that x has to exist, except by this definition. Thus, we must be shown *why* this definition is correct before we accept it.

Pertsel Vladimir from the Weizmann Institute of Science's Computation Center, comments:

229

Goedel's
Mathematical
Proof of the
Existence of
God

> I've heard the following advanced as a proof of God's existence:
>
> 1 "God" is defined as having all qualities in perfection.
> 2 Existence is a quality.
> 3 Therefore God must have this quality of existence, and so must exist.

However, the template for disproving such reasonings was given by A. G. Gein at one of the All-Soviet-Union mathematical competitions:

> 1 Non-existence is a quality.
> 2 Therefore God must have this quality of non-existence.
> 3 Therefore God must not exist.

Kazimir Kylheku from the Computer Science Department of the University of British Columbia, Vancouver comments:

> Maybe it was just happy hour in Austria, and Goedel was entertaining a less than erudite gathering.

From an anonymous graduate student in Electrical and Computer Engineering at Rice University in Houston, Texas:

> Pascal's "Pensées" ("Thoughts") came about because the Church urged him to make some public comment on religion. The following may contain the "ontological argu-ment" for the existence of God:
>
> 1 Suppose God does not exist.
> 2 Imagine the most beautiful and perfect place imaginable.
> 3 Imagine (2) with God. It is more beautiful and perfect than (2) without God.
> 4 (3) contradicts the supposition that (2) is the most perfect place imaginable.
> 5 Hence assumption (1) is incorrect.
>
> This five-step argument is inadequate because of assumptions about the nature and limits of "imagination." I have generally found "proving" the "existence" of God to be a shallow pursuit. It is much more interesting to lessen the emphasis on "proof" and increase the subtlety of what is meant by "exist." For example, I believe that my sensible world has incredible order that implies plan and causation. It also has disorder and power which makes me feel weak and small. It is soothing to accept both of these and let the sheer wonderfulness of it comfort my existential worries. When I have an epiphany that the humor in the pose of a cat on a 4000-year-old baked clay sculpture connects, me and the cat in my garage to a person from another time and place, and I open myself to that moment and let it reaffirm my respect for humanity, is that a prayer? There is no question in my mind that Unity is apparent. Some would argue that it is constructed in my mind like a Rorschach test. So what, let them be dour and determinis-tic. It won't stop me from stealing those moments of pure joy when I can! For more on this line of reasoning, read Joseph Campbell's *The Inner Reaches of Outer Space* and Walker Percy's *The Last Self-Help Book in the Cosmos*. The former is short and rigorous with arguments based on semiotics, and the latter is fun, yet profound. People are trained to look for "God" and "religion" outside themselves. Even when they are told that "God is in each of us," the problem is just that: they are "told."
> On another topic, Walker Percy, in his book *The Last Self Help Book*, provides a

structure by which art can be considered a God-substitute in its role as a conduit between ourselves and the Universal/Infinite. The appeal of math for artists is precisely the universality of mathematics—universal in the sense of culturally invariant. This is not mathematics as a tool to prove God but rather mathematics as evidence of some universal fabric for existence. Perhaps these artists are unaware that all of mathematics is prefixed by the same post-modernist-like disclaimer (authored by Goedel, Russell, and Whitehead) that appears in current art/anthropology/cultural criticism. On the other hand, perhaps they *are* aware that mathematics has no absolute claim, but nonetheless joins hands with most practicing mathematicians in a willing suspension of disbelief.

Mark Hopkins, a consultant from Milwaukee, comments:

I. Robinson's Logic. I've been considering another proof, somewhat similar in nature, that starts with Robinson's investigations into idealizations. (Abraham Robinson (1918-1974) was a logician and mathematician famous for his "non-standard analysis.") Robinson established a mode of inference whereby one could perform the following type of deduction: *for all x there exists y such that ($x \ll y$) yields: "there exists y* such that for all x ($x \ll y$*)."* The expression y* is called an ideal object, and constraints are imposed on the nature of the relation $\ll$. My proof seeks to start out with a property that simultaneously identifies an essential aspect of God and satisfies the constraints Robinson placed. As an example (which doesn't actually work) consider "x is loved by y." The Prime Mover argument would use the relation "x is caused by y," but the problem here (apart from the issue of what causality is) is that there may only be a finite number of instances of causal links in the universe. If I understand Robinson correctly, you need an infinite relation. This gets to the heart of the matter.

II. Definition Problem. Here's another point to consider. People have apparently lost their common sense when studying proofs of God, and have completely forgotten the idea of *Definition Before Proof.* One has to *define* what it is meant by God before trying to prove existence. Therefore, the lack of resolution over the issue has absolutely nothing to do with the supposed inapplicability of logic here, but rather in the lack of consensus over what constitutes a valid definition.

III. Nobody Can be an Atheist. This discussion is the basis for why I argue that *nobody can be an atheist (or a theist).* The reason is that in order to be either, you have to know what it is you're (dis)believing in first. You can't disbelieve something if you don't even know what it is you're not believing in. This would be dishonest. Likewise, you can't believe in something if you don't even know what it is you're believing in—that's delusion (actually, it's idolatry). This makes everyone an agnostic until or unless a definition is agreed upon.

Dr. George Markowsky from the University of Maine's Computer Science Department comments:

I did not find Goedel's proof very convincing or particularly interesting. Is there good evidence that he put this forward in a serious manner? Perhaps he was just fooling around. One of your respondents suggested that we had to be agnostics rather than atheists. A. J. Ayer, in his book *Language, Truth and Logic* (Dover, 1946), argues for a fourth position on God. He claims that the standard three positions—believer, atheist, and agnostic—don't make sense because the concept of God is not well defined. Ayer argues it makes no sense to state a position about God's existence since the concept makes no sense.

Arturo Magidin is a second year graduate student in the mathematics Ph.D. program at the University of Berkeley. He comments:

231

Goedel's
Mathematical
Proof of the
Existence of
God

1) The first problem arises in Axiom 2 (Closure). What does "necessarily contains" mean? When do we say that a property contains another? If it implies it? Then since $a \to a$ is always true, you haven't really said anything. Note also if you define any property to be positive (and negative), then axioms 1 and 2 are satisfied. 2) The proof up to Axiom 3 is "proving by definition." It is analogous to defining a "Real Loch Ness Monster" as a huge animal that lives in the Loch Ness and that actually exists. Then I claim that a Real Loch Ness Monster necessarily exists. This is indeed true, since I have, in effect, defined a Real Loch Ness monster to exist. However, the possibility that there is no such thing as a Real Loch Ness Monster still holds; i.e., you cannot define things into existence simply by stating that they have the property of existing. 3) In Axiom 4, the term "minimal" is used. Minimal with respect to what? (This is not a critique; this is an honest question; I think I might be missing something.) Theorem 2 also appears to be "defining things into existence." Finally, by the time we have finished the proof at Theorem 3, it seems that taking every property to be both positive and negative, you have satisfied all axioms. So, if x is God-like, it necessarily exists, but that is not the same as saying that there necessarily exists something which happens to be God-like. You have in essence, paraphrased Descartes' argument:

Descartes Proof Of God's Existence
God is a being which by definition possesses all positive properties. Existing is a positive property; therefore, God has the property of existing; hence God exists.

This argument has already been demolished by philosophers. I recommend a perusal of Michael Martin's *Atheism: A Philosophical Justification* for a clearer explanation of the "real Loch Ness monster" scenario.

Benjamin J. Tilly, a mathematics graduate student at Dartmouth, comments:

Goedel's proof of God's existence seems to have a number of flaws. The main one is that it asserts a number of axioms which are not very reasonable, and the second one is that there is no guarantee that the resulting "God" has any relationship to the ordinary religious notions. 1) Axiom 1 gives a definition. It is fine. However, I would like to know what a "property" is, and what "negative" means. 2) Axiom 2 seems to be missing something since if I define all properties to be both negative and positive then the two axioms are satisfied and the theorem is false. 3) Does the definition, "Something is God-like if and only if it possesses all positive properties" have any relationship to the usual notions of God? 4) I have no problem with Axioms 3 and 4. 5) What does "minimal" mean? Minimal in what sense? 6) Theorem 2, followed by the definition and Axiom 5, essentially asserts the existence of God, as an axiom. Of course if you want to do that then I could just assert that God exists.

Mats Andtbacka, a chemical engineering student from Finland, comments:

All right, so I'm not a logician. But if a property <A> is defined as being something <B> *if and only if* something else <C>, and then in the next sentence you say that a property <D> is also <B>, provided merely that it contains an instance of <A or D>—well, isn't that a problem? Or is my expression of this the problem here? Heck, who cares … So there necessarily exists at least one x such that x possesses all positive properties, where a property is positive if its negation is negative. I'm not sure I'd be impressed even if I understood that.

Geoffrey Moore from the University of Maryland comments:

1) Axiom 1 relies on a two way street. What if I said "I am lying"? This is the first paradox which states that Axiom 1 applies to absolutely nothing. 2) In Theorem 2,

Goedel is being stupid. He says that a unicorn exists because horns exist. Wrong!
3) Finally look at the definition, "A property P is the essence of x if and only if x has P and P is necessarily minimal." Is this a circular argument or what? x has P has x? Wrong!

I asked Scott Pallack, a taxi driver in Fort Lauderdale, Florida, "Is there anything you personally would consider mathematical evidence of the existence of a deity?" He responded:

> One cannot prove that an omnipotent God does not exist. (Assume you have such a proof. You cannot prove that God has interfered with your mind making you think the proof is valid when it is not.) Of course, this does not prove the existence of God. Personally, I think the question is undecidable in the Goedelian sense. You can assume either G or not-G as an axiom. Just be prepared to accept the consequences of your assumption.
>
> Pascal had a so-called "proof" of God's existence which contains a statement similar to, "Even if there is a tiny probability that belief in God sends you to heaven and disbelief sends you to hell, one should believe in God, because there's nothing to lose." However, this reasoning has a flaw. For example, consider the possibility that only believers go to hell.

Bert Smits is a doctoral student working on dynamical systems in the Limburgs Universitair Centrum, Diepenbeek. He comments:

> I've attended a talk by a mathematician from Laval University, Canada, who lectured in Europe on "The Existence of the Uncaused Cause." If you define God to be the uncaused cause, then I think he had a rather convincing proof of God's existence.

Goran Wicklund works for a subsidiary to Ericsson Telecom as a system designer of switching equipment. He comments:

> Some mathematicians consider $e^{i\pi} + 1 = 0$ to be "God's formula." It contains the three most important mathematical terms: e, π, and i. It also contains the additive, multiplicative, and exponentiation operator and the two neutral units for these operators (0 and 1). Some believe that this compact formula uniting various facets of mathematics is surely proof of a Creator.

Allan Trojan, an Associate Professor of Mathematics from Atkinson College, York University, comments:

> The idea of a mathematical proof of the existence of God does not interest me in the slightest. God is an "energy." You don't prove or disprove the existence of an energy, you use it. I am a mathematician (not terribly renowned however), and I am religious. I am a mystic.

St. Anselm (1033–1109) was the Archbishop of Canterbury, and a philosopher famous for his arguments for the existence of God. Langley Muir, a mathematical logician at the Natural Resources Canada, Ottawa, comments:

> There is a proof that was first postulated, perhaps by St. Anselm, known as the Ontological argument. It has a very long history in philosophy and is one of the major arguments used by Descartes to prove the existence of God. Shorn of a lot of supporting arguments, it is a simple syllogism in "Barbara." (Muir explains the term "Barbara" in Appendix 2 of this chapter.)

233

Goedel's
Mathematical
Proof of the
Existence of
God

1   Major premise: God is perfect
2   Minor premise: Perfection exists
3   Conclusion: God exists

There is a long list of books and articles related to this argument which come close to a real mathematical proof, if you accept that logic and mathematics are the same thing. For more information, see: Plantinga, A. (1965) *The Ontological Argument: From St. Anselm to Contemporary Philosophers*, Doubleday-Anchor: New York.

Tim Poston is a researcher at the Institute of Systems Science at the National University of Singapore where he is a principal investigator of a heart project dealing with medical imaging and three-dimensional graphics. He has also published in physics, economics, archeology, and computer vision. He comments on the previous example of St. Anselm's proof:

> The ontological argument is a bit more subtle than explained. It defines God (in modern terminology) as a maximal element in the lattice of good entities, so it needs Zorn's Lemma to succeed. Thus it depends on the Axiom of Choice. As pointed out in the journal *Manifold* around 1970, this makes the existence of God *equivalent* to the Axiom of Choice, since an omnipotent deity could clearly do the choosing.
>
> I've often thought that modern mathematics could contribute a lot of pizzazz to numerical theology. The numerology of the religious depends on facts known since Pythagoras. For example, observe that since, according to Revelation, the saints are "a multitude that no man can number," most saints are clearly not descendants of Adam, who by a simple generation-labelling argument, can be at most a denumerable set, even if Armageddon is [countably] infinitely deferred. Religious reasons do not excuse violence: they accuse religion.
>
> I recommend the following article: "Ontology Revisited" by Vox Fisher, *Manifold* 6, March 1970. The countability point regarding saints was an exercise in my piece "Aquinas' God versus Maxwell's Demon" in *Manifold* 11.

Paul Swingle, a computer programmer from Boca Raton, Florida, comments:

> The responses describing Anselm's proof of the existence of God remind me of the high point of my philosophical career in school. We were discussing Anselm's proof, which as I recall was:
>
> 1   God is defined as, "That than which there is nothing more perfect."
> 2   That which exists is more perfect than that which does not exist.
> 3   Therefore God must exist.
>
> From this I created Swingle's Proof of the Existence of the Devil:
>
> 1   The Devil is defined as, "That than which there is nothing more evil."
> 2   That which exists is more evil than that which does not exist.
> 3   Therefore the Devil must exist.
>
> I think this rests on the notion of existence being more "something" than nonexistence. A perfect triangle cannot exist; the notion must be debased or harmed in some way by having a concrete example.

David Wilson of Cabletron Systems Inc. comments:

> Mathematics effectively amounts to the formal manipulation of symbols. Inasmuch as an idea can be couched in terms of symbols amenable to such formal manipulation, that

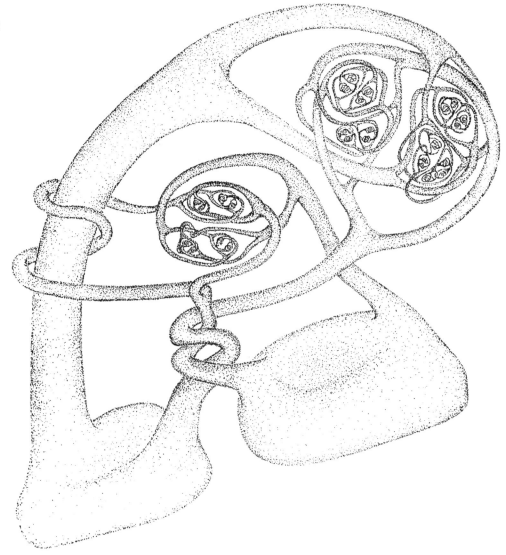

Mating Dance of the Horned Spheres, a visual proof of the existence of God by Tim Poston.

idea is mathematical, inasmuch as it cannot, it is outside the realm of mathematics. The power of mathematics is derived from the vast store of symbolic manipulations built up over the centuries. In order to benefit from this power, other disciplines often strive to translate their non-symbolic ideas into mathematical symbols, with varying degrees of success. In most cases, the translation is an approximation; in no case is such a translation provably correct. The concepts from many areas of discourse, such as religion, philosophy, social sciences, economics, and politics are at best poorly translated into mathematical symbols, and consequently not amenable to sound mathematical argumentation. Just as within a discipline closely allied to mathematics, such as physics, it would be laughable and demeaning to ask for a mathematical proof of the existence of an electron. Similarly a proof of the existence/nonexistence of God is all the more laughable and demeaning. A mathematical proof of any idea presupposes that that idea

235

Goedel's
Mathematical
Proof of the
Existence of
God

can be couched in symbols, which is generally not the case. I, for one, side with Lebesgue on this subject: "In my opinion, a mathematician need not preoccupy himself with philosophy." (Lesbesgue, as quoted in *The Foundations of Geometry and the NonEuclidean Plane*, by George E. Martin)

David Joyce from Clark University's Department of Mathematics and Computer Science comments:

> All of this reminds me of an equally circular proof of God. Define God as a perfect being, posit that perfection includes existence, conclude existence of God.

John Baez, a mathematical physicist from the Mathematics Department at the University of California, Riverside, comments:

> Here is an old proof of God's existence. Consider: "*If this sentence is true, then God exists.*" Call that sentence P for short. Assume for the moment that P is true. Then, since P *says* if it's true then God exists, God exists. Of course, we just *assumed* P was true; we don't know this. However, look what we have shown: *if* P is true, then God exists. But that's just what P says! So we have shown P is true! Therefore God exists. QED.
>
> Oddly enough, this proof works equally well if we change "exists" to "does not exist." Therefore, the proof is very versatile. Usually I use it to prove that I don't need to get up in the morning. For more, try reading about provability predicates and Loeb's theorem in Boolos and Jeffrey's *Computability and Logic*, although somehow I suspect Smullyan's books should have something to say about this one. Sentences that say "this sentence implies this sentence is provable" have an amusing tendency to be provable by an argument similar to the above one.

Tony Davie, a lecturer in computer science specializing in functional programming and programming languages at St. Andrews University (Scotland), makes a comment similar to the previous one:

> The following argument, perhaps due to Smullyan, is amusing. Consider the self-referential statement "If this statement is true then God exists." Let us try to prove it:
>
> 1 The statement is true.
> 2 If the statement is true then God exists.
> 3 God exists.
>
> So, by the deduction theorem, the hypothesis 1 implies 3, i.e. if 1 then 3. If the statement is true then God exists. Thus the statement is true thus God exists.

Joshua Weiner from Oberlin College's Department of Physics comments:

> There is a problem with the Goedel's proof of God. "Existing" is not necessarily better than not existing. Consider the field of quantum mechanics. A particle does not exist until observed. *Does this make us gods for being able to turn a probability field into a particle?* Similarly, by existing and encompassing all, a god must encompass evil as it is better to have all than to have things outside of one's self. Yet this would contradict the definition of god as being all good. Also there is the quality of being weak and impotent. If god is all encompassing it should have all qualities including weakness, stupidity, anger, hate, and lust. To have less would to be less than perfect. God is a logical contradiction. This isn't a mathematical proof, it's just an new wording of an old "proof."

From an anonymous Internet traveler, software engineer, and mathematician named Sue:

1) Good grief, anyone can make up an axiom like Axiom 3 in Goedel's proof of God. Maybe being God-like is really a negative property (negative and positive not having been defined in the first place). 2) The meaning of Axiom 4 is unclear. 3) I am not sure how Theorem 2 follows from the definition before it. 4) Regarding Theorem 3, is there a limit on how many such x's exist? This is a joke, right? This whole proof depends on a set of axioms that I find suspect, to say the least. And the definitions don't seem to define anything that actually exists. But some poor [expletive deleted] is going to fall for it, and go around saying, "I know God exists, because this really smart guy proved it."

Joe McCauley, a programmer from Colorado Springs, Colorado, comments:

Being both a religious person and having an analytical personality, I have struggled with the question of how I can really know that God exists. I tend to regard proofs in this chapter as nothing more than clever logical manipulations of words and symbols that provide no real information on the question of whether or not God truly exists. What I would like to see is physical evidence, or something which could be demonstrated experimentally. I do not expect to find this.

After reading some of the works of John Powell (a Jesuit priest at Loyola University), I concluded that it is not possible to know the existence of God in the sense of being able to show a rigorous proof or experimental evidence of God's existence. Belief in God requires some degree of faith; for those who refuse to accept faith, there will always be room to challenge any alleged proof of God's existence, and to doubt any alleged divine intervention in anything they observe. And those who have faith don't need the sort of proof being discussed here. You might say I believe in God because I choose to, and you'd probably be right.

Dan Platt, an IBM physicist with an interest in religion, comments:

Goedel is spinning wheels. One can define things, postulate axioms about them, and prove theorems about them. The assignment of words such as "God-like" provides a context or connotation to these things. However, it isn't clear that every theologian would care to describe God in this kind of language. Further, I've heard this argument before in slightly different language. There's possibly a need for the axiom of choice in this (you don't know what the cardinality of positive properties is). Maybe Goedel is assuming a countable or even finite set.

Timothy Chow, a doctoral student from MIT, comments:

Let Sentence A be the sentence, "If Sentence A is true, then Santa Claus exists." If we replace truth by theoremhood, can we get something along the lines of Goedel's theorem by trying to formalize this paradox?

Torkel Franzen, a researcher and philosopher at the Swedish Institute of Computer Science, Kista, comments on Henkin sentences. (Henkin sentences of the form "This sentence is provable," are automatically true. The key is the relation to the Santa Claus paradox.)

Why are Henkin sentences self-proving? The reasons follows from *Loeb's theorem:* if the formula "if A is provable in T, then A" is provable in T (for T satisfying certain conditions), then A is provable in T. As far as I know there isn't any special simpler argument that applies to Henkin sentences in particular.

There are two proofs of Loeb's theorem. First there is Loeb's original simple proof which deserves to be well known in wider circles as Goedel's proof. First note that Loeb's theorem implies Goedel's second incompleteness theorem as a special case: taking A to be the formula $0 = 1$, we conclude from Loeb's theorem that the consistency

237

Goedel's
Mathematical
Proof of the
Existence of
God

of T is not provable in T unless T is inconsistent. So the reasoning involves some prooftheoretical aspects absent in the proof of Goedel's first theorem. Understanding the idea of the proof is just as easy as understanding the idea of using the Liar paradox to obtain Goedel's first theorem. Loeb starts from a variant of the Liar. Where the Liar is "this sentence is false" and the paradoxical reasoning draws a contradictory conclusion, the present paradox, which we may call the Santa Claus paradox, purports to prove the existence of Santa Claus. Here is the argument: Let p be the statement "if p is true then Santa Claus exists." Now, if p is true, then if p is true, Santa Claus exists. This means that if p is true, Santa Claus exists. But this is just what p says. So p is in fact true. But then, since if p is true, Santa Claus exists, it follows that Santa Claus exists.

To transform this reasoning into a valid argument in T, we need a provability predicate satisfying certain conditions. So suppose we have a formula Bev(x), formalizing "x is the Goedel number of a variable-free formula provable in T." We assume that Bev(x) satisfies the following conditions, where C stands for the Goedel number of a formula C. (I) If C is provable in T, Bev(C) is provable in T. (II) Bev($A \to B$) $\to$ (Bev(A) $\to$ Bev(B)) is provable in T for all A,B. (III) Bev(A) $\to$ Bev($Bev($A$)$) is provable in T for all A.

Now suppose the formula A satisfies the condition that the formula Bev(A) $\to$ A is provable in T. In other words, it is provable in T that if A is provable in T, A is true. Now, using a mechanism for self-reference as in Goedel's proof, we obtain a formula B for which it is provable in T that (2) B $AL \to$ (Bev(B) $\to$ A). We now show that A is provable in T, by the following reasoning. From (2) we get that it is provable in T that (3) B $\to$ (Bev(B) $\to$ A) and hence, using principles (I) and (II) we get that it is provable in T that (4) Bev(B) $\to$ Bev($Bev($B$) $\to$ A$) and so by another application of (II) that it is provable in T that (5) Bev(B) $\to$ (Bev($Bev($B$)$) $\to$ Bev(A)). From (III) and (5) we get that it is provable in T that Bev(B) $\to$ Bev(A). But (6) and (1) yield that it is provable in T that (7) Bev(B) AR A By (2) and (7) B is provable in T, so by (I), Bev(B) is provable in T. Hence, by (7), A is provable in T.

Now if you work through this, you'll see that it mimics the reasoning in the Santa Claus paradox. A funny thing about Loeb's theorem is that it went unnoticed for years that it actually follows immediately from Goedel's second theorem. For assume (1) holds but A is not provable in T. Then the theory T + ~ A is consistent. But in the theory T + ~ A, it is provable that A is not provable in T, i.e., that T + ~ A is consistent. So, by Goedel's second theorem, T + ~ A is inconsistent. Hence A is provable in T. Now I like the original proof much better, since it extends the proof of Goedel's second theorem in a natural way using a refined version of the Liar. Also, the original argument applies to intuitionistic theories as well as classical ones, whereas the second proof applied to an intuitionistic system only shows that ~ ~ A is provable in T.

I recommend a quite readable chapter "The Incompleteness Theorems," by C. Smorynski, in the *Handbook of Mathematical Logic* (North-Holland Publishing Company, 1977), edited by Jon Barwise, which contains fascinating information for those who are interested in the subject. See also Smullyan's books.

Randall Holmes is a professor of mathematics at Boise State University in Boise, Idaho. He is a mathematical logician by training. He teaches and does research on Quine's set theory "New Foundations" and on computer-assisted reasoning. He comments:

If we try to work by analogy with Goedel, we build a one-place predicate of natural numbers which says, "If one-place predicate *n* is provable of the number *n*, then [Santa Claus exists]." This sentence has a Goedel number; let this number be G. Predicate number G holds of G exactly if "if one-place predicate G is provable of the number G, then [Santa Claus exists]." Thus, we have a sentence which says, "If I am provable, then

Santa Claus exists." If Santa Claus exists, this statement is true; it is provable if the existence of Santa Claus is provable, and not provable otherwise. If Santa Claus does not exist, this statement is true but unprovable just as in the case of Goedel's sentence which asserts that it itself is not provable. Nothing is added to the force of Goedel's argument by this construction. The point is that the Santa Claus paradox is really exactly the paradox of the Liar ("This statement is false"), and so its translation into the realm of provability is exactly Goedel's diagonalization.

MANIFOLD AND MATHEMATICAL THEOLOGY

> *Most theology is more about particular human grammars than about universal truth. "I think, therefore I am" is perfectly paralleled by, "It is raining, therefore it exists," which clearly tells us nothing at all about "it."*
>
> —Tim Poston, 1994

Manifold—the only journal ever published entirely on the subject of mathematical theology—began as a publication created by half a dozen graduate students at the University of Warwick (UK). The journal started in 1968 and ran for 20 issues, until 1980. Tim Poston, a researcher at the Institute of Systems Science at the National University of Singapore, led me to several interesting articles regarding God and mathematics which appeared in *Manifold*.

Ontology Revisited

> *Most theology is word games ... getting out of language what was earlier snuck into it; the same is of course true about mathematics.*
>
> —Tim Poston, 1994

In *Manifold* 6 (1970, pp: 48–49), Vox Fisher published "Ontology Revisited," a fascinating paper on a mathematical proof of God's existence. The paper starts:

Theorem: The axiom of choice is equivalent to the existence of a unique God. (St. Anselm, Aquinas, and others.)

Proof: Partially order the set of subsets of the set of all properties of objects by inclusion. This set has maximal elements. God is by definition (due to Anselm) a maximal element set:

God $\subseteq$ God $\cup$ {existence}, so God = God $\cup$ {existence} $\therefore$ God exists.
Uniqueness: if God and God′ are two gods, then God $\cup$ God′ $\supseteq$ God (due to Aquinas), and $\therefore$ God $\cup$ God′ = God, $\therefore$ God $\subseteq$ God′, and similarly God′ $\subseteq$ God

$$\therefore \text{ God} = \text{God}'.$$

Given a set $\{A_\alpha\}_{\alpha \in R}$ of sets, let the unique God pick $x_\alpha \in A_\alpha$ for each $\alpha \in A$. (He can do so by omnipotence.) Then $(x_\alpha)_{\alpha \in R} \in \prod_{\alpha \in R} A_\alpha$ as required.

Aquinas' God versus Maxwell's Demon

239

Goedel's
Mathematical
Proof of the
Existence of
God

There cannot be design without a designer, order without choice ... Arrangement, disposition of parts, and subservience of means to an end imply the presence of intelligence and mind.

—William Paley

In *Manifold* 11 (1971, pp: 71–72), Tim Poston published "Aquinas' God versus Maxwell's Demon," an unusual paper on thermodynamics and God. As background to the arguments in his paper, the Second Law of Thermodynamics states that in any isolated system, entropy (disorder) must always increase. (The term "isolated" is important because entropy can decrease locally at the expense of greater increase elsewhere.) The Second Law is evident in the shuffling of cards: it is easier to randomly shuffle cards *out* of a particular sequence than into it. As the cards are shuffled, information is lost.

The following is excerpted with permission from Poston's paper which attempts to show that God's omniscience implies the Second Law of Thermodynamics:

We now return to mathematical theology, as follows:

1 God is omniscient.
2 Hence, He knows the position of every particle at every time.
3 Hence, He is observing every particle at every time. (Indeed, particle existence depends upon this observation, cf. the writings of Bishop Berkeley.)
4 Hence, He is gaining information about all parts of the Universe at all times and in all places.
5 Hence, he is increasing the entropy of the Universe in all times and in all places.
6 Hence, the entropy of the Universe increases in all times and in all places. (This is the Second Law of Thermodynamics. Q.E.D.)

Corollary 1. The Axiom of Choice implies the Second Law of Thermodynamics.

Corollary 2. God can raise the dead. Proof: Death and decay involve increase of entropy. God can reverse these processes by means of sacrifice of some of His immense store, accumulated as above, of confidential information.

Corollary 3. God will die. Proof: The Universe will end with a war producing chaos (The Revelation of St. John the Divine), after which God will put it together again as chaos, after which God will put it together again as a whole using all his information. It will then remain perfect (in contrast to the present state). Hence God will cease to observe it, hence He will cease to satisfy the definition of Himself (*Summa Contra Gentilla*, St. Thomas Aquinas), hence He will cease to exist.

Sanctification and the Hopf Bifurcation

In *Manifold* 19 (1977, pp: 25–26), Phillip Holmes published the article, "Sanctification and the Hopf Bifurcation," a thought-provoking paper on the halo (the circle of light surrounding the heads of saints as a symbol of holiness and sanctification). In his paper, Holmes uses chaos theory to suggest that "sanctification consists of the creation of an attracting closed orbit in Hopf bifurcation."

APPENDIX I OF POSTSCRIPT 1: ST. AUGUSTINE AND THE "CITY OF GOD"

Atheist: *"I don't believe in God."*

Abdu'l-Baha: *"I don't believe in the God that you don't believe in."*

Interestingly, it seems that the existence of the set of all natural numbers is sanctioned by one of the Fathers of the Church. For example, St. Augustine (see Chapter 6) often speaks of the relationship between numbers and God in his book *The City of God*. The following is from Book XII, Chapter 19.

The answer to the allegation that even God's knowledge cannot embrace an infinity of things:

Then there is the assertion that even God's foreknowledge cannot embrace things which are infinite. If men say this, it only remains for them to plunge into the depths of blasphemy by daring to allege that God does not know all numbers. It is certainly true that numbers are infinite. If you think to make an end with any number, then that number can be increased by the addition of one. More than that, however large it is, however great the quantity it expresses, it can be doubled; in fact, it can be multiplied by any number, according to the very principle and science of numbers.

Every number is defined by its own unique character, so that no number is equal to any other. They are all unequal to one another and different, and the individual numbers are finite, but as a class they are infinite. Does that mean that God does not know all numbers, because of their infinity? Does God's knowledge extend as far as a certain sum and end there? No one could be insane enough to say that.

Now those philosophers who revere the authority of Plato will not dare to despise numbers and say that they are irrelevant to God's knowledge. For Plato emphasizes that God constructed the world by the use of numbers, while we have the authority of Scripture, where God is thus addressed, "You have set in order all things by number, measure, and weight." And the prophet says of God, "He produces the world according to number"; and the Savior says in the Gospel, "Your hairs are all numbered."

Never let us doubt, then, that every number is known to him "whose understanding cannot be numbered." Although the infinite series of numbers cannot be numbered, this infinity of numbers is not outside the comprehension of him "whose understanding cannot be numbered." And so, if what is comprehended in knowledge is bounded within the embrace of that knowledge, and thus is finite, it must follow that every infinity is, in a way we cannot express, made finite to God, because it cannot be beyond the embrace of his knowledge.

APPENDIX II OF POSTSCRIPT 1: BARBARA

Langley Muir used the term "Barbara" in his Internet response. "Barbara" refers to the medieval classification of syllogisms. A syllogism is a deductive scheme of a formal argument consisting of a major and minor premise and a conclusion. Example: Every virtue is laudable; kindness is a virtue; therefore kindness is laudable. An odd example from Bob Stong: Hard work is good for you. Stealing refrigerators is hard work. Therefore, stealing refrigerators is good for you. Langley Muir notes that the syllogistic logic allows four types of proposition: 1) A (universal affirmative) subject distributed, predicate undistributed. Example: All men are mortal. 2) E (Universal negative) subject distributed, predicate distributed. 3) I (particular affirmative) subject undistributed, predicate undistributed. 4) O (particular negative) subject undistributed, predicate distributed.

Using these four types of propositions and some rules of argument, all of the valid forms of syllogism may be derived. To remember what they are, one of the more famous sets of memory-lines was given for the first time by Pope John XXI (died 1277): Barbara, Celarent, Darii, Ferioque, prioris; Cesare, Camestres, Festino, Baroko, secundae; tertia, Darapti, Disamis, Datisi, Felapton, Dokardo, Ferison, habet; Bramantip, Camenes, Dimaris, Fesapo, Fresison, quarta insuper addit.

The first, second, third, and fourth refer to the four moods, and the nouns (in upper case) each have three vowels which indicate the form of the propositions making up a valid syllogism, consists of a major premise (first), a minor premise (second), and a conclusion (third). Hence a syllogism is Barbara if a syllogism is made up of major premise A, a minor premise, and a conclusion all using propositions of the form A. Whence: "God is perfect; perfection exists; therefore, God exists" is a syllogism in Barbara. All of this is found in any good book on syllogistic logic that would be used in a first or second year undergraduate philosophy course on logic. (See, for example: Eaton (1959) *General Logic: An Introductory Survey*, Charles Scribner's Sons: New York.)

241

Goedel's
Mathematical
Proof of the
Existence of
God

Mathematicians Who Were Religious

It is completely wrong for people to assume that a true scientist cannot simultaneously be a true man of God, believing in God as Creator and Savior and believing the Bible as God's revelation.
—Henry Morris, *Men of Science, Men of God*, 1982

The theory of limits, on which modern mathematics, and as a corollary, all modern science and technology depend, is actually a secularized form of the Scriptural view of reality. Mathematics and Scripture both view the human grasp of reality as increasingly approximated to, but never identified with, the human symbols we use to represent reality.
—Ford Lewis Battles, 1978

Over the years, many of my readers have assumed that famous mathematicians cannot be religious. In actuality, a number of important mathematicians were quite religious. As an interesting exercise, I conducted an Internet survey where I asked respondents to name important mathematicians who were also religious. The following list is sorted in order of the number of "votes" each scientist received. For example, Isaac Newton and Blaise Pascal were the most commonly cited religious mathematicians.

The Top 22

1   *Blaise Pascal* (1623–1662) French geometer, probabilist, physicist, philosopher, and combinatorist. Inventor of the first calculating machine. A deeply spiritual man and a leader of the Jansenist sect, a Calvinistic quasi-Protestant group within the Catholic Church. He believed that no one loses who chooses to become a Christian. If the person dies, and there is no God, the person loses nothing. If

there is a God, then the person has gained heaven whereas his skeptical friends will have lost everything in hell. One of my Internet respondents writes:

> Pascal in his early childhood sought to prove the existence of God. Since Pascal could not simply command God to show himself, he attempted to prove the existence of a devil so that he could then infer the existence of God. He drew a pentagram on the ground and then became so afraid that he ran away. Pascal said that this experience made him certain of God's existence.

2 *Isaac Newton* (1642–1727) English mathematician, physicist, and astronomer. Invented calculus. Law of Gravitation. Author of many books on biblical subjects, especially prophecy. Creationist. Several respondents noted that Newton wanted to be known more for his theological writings than for his scientific/ mathematical writings. Other respondents said that Newton believed in a Christian unity as opposed to a trinity. "Isaac Newton developed calculus as a means of describing motion, and perhaps for understanding the nature of God through observation of nature."

3 *Leonhard Euler* (1707–1783) Swiss mathematician and the most prolific mathematician in history. Son of a reformed vicar. Leonhard Euler is said to have been quite distressed at being unable to prove mathematically the existence of God.

4 *Rene Descartes* (1569–1650) Philosopher and mathematician. Born in France. Founded analytic geometry. Metaphysics.

5 *Marin Mersenne* (1588–1648) French theologian, philosopher, and number theorist. Priest and monk.

6 *Georg Boole* (1815–1864) British logician and algebraist. A vicar.

7 *Donald Knuth* (1938–) Computer scientist and mathematician. "One of the greatest living computer scientists." An active Lutheran and Sunday School teacher. One respondent writes to me:

> Knuth wrote a beautiful book titled *3:16*. The book consists entirely of commentary on Chapter 3, Verse 16 of each of the books in the Bible. (If Chapter 3 of a book does not have 16 Verses, he examines Chapter 4. If the book is too short for this trick to work, he omits that book.) Knuth also includes calligraphic renderings of the verses. Evidently from his work with word-processing and fonts, Knuth developed international acquaintances with calligraphers.

8 *Srinivasa Ramanujan* (1887–1920) Indian number theorist and devout Hindu. Ramanujan often believed that his mathematical insights were brought to him in dreams by a deity worshipped by his family.

9 *Louis Augustin Cauchy* (1789–1857) Very devout Catholic, French analyst, applied mathematician, and group theorist. Cauchy is second to Euler in being the most prolific mathematician in history.

10 *Georg Freidrich Bernhard Riemann* (1826–1866) German mathematic who made important contributions to geometry, number theory, topology, mathematical physics, and the theory of complex variables. Non-Euclidean geometry. He attempted to write a mathematical proof of the truth of the book of Genesis. A student of theology and biblical Hebrew. The son of a Lutheran minister.

11 *Kurt Goedel* (1906–1978) Czechoslovakian-American logician, mathematician, and philosopher. One of my respondents writes:

> I must admit to curiosity about Goedel. After Russell referred to Goedel as being Jewish, somebody wrote to Goedel about it. Goedel said he was not Jewish, but attached no importance to this issue. Also, in the preface to Goedel's collected works, there's a passing mention of some notes on demonology.

Others have noted that Goedel was fascinated by the afterlife and the existence of God. Goedel thought it was possible to show the logical necessity for life after death and the existence of God. In four long letters to his mother, Goedel gave reasons for believing in a next world.

12 *Gottfried Wilhelm von Leibniz* (1646–1716) German analyst, combinatorist, logician, and co-inventor of calculus. Inventor of a mechanical multiplication machine. Argued for the existence of God.

13 *Josef Maria Wronski* (1778–1853) Polish-born analyst, philosopher, combinatorialist, and physicist. Inventor of the Wronskian, an important concept in linear algebra and differential equations. An ardent seeker of mathematical explanations for history and religion.

14 *Georg Cantor* (1845–1918) German set theorist with an interest in the infinite. A devout Lutheran.

> Cantor felt a duty to keep on, in the face of adversity, to bring the insights he had been given as God's messenger to mathematicians everywhere. (Dauben, J., *Georg Cantor*, Princeton University Press, 1990, p. 291)

15 *Pierre Simon Laplace* (1749–1827) French analyst, probabilist, astronomer, and physicist. Best known for his work on celestial mechanics, probability theory, and differential equations.

16 *Luitzen Egbertus Jan Brouwer* (1881–1966) Dutch topologist and logician. A member of the Dutch reformed church. "Brouwer had very peculiar thoughts about church and religion."

17 *Leopold Kronecker* (1823–1891) German algebraist and number theorist. Kronecker said, "God created the natural numbers, and all the rest is the work of man."

18 *Charles Babbage* (1792–1871) English analyst, statistician, and inventor. Prophet of the modern computer. Author of the ninth and last *Bridgewater Treatises*, including a mathematical analysis of the biblical miracles.

19 *Isaac Barrow* (1630–1677) English theologian, geometer, and analyst. Barrow is best known as the teacher of Newton; however, Barrow was also a talented mathematician.

20 *Alfred North Whitehead* (1861–1947) English algebraist, logician, and philosopher.

21 *Rev. Charles Dodgson* (1832–1898) Respected mathematician, best known today as Lewis Carroll, author of *Alice in Wonderland* and *Through the Looking Glass*, both of which contain references to logic and theology.

22 *John Harris* (1666–1719) English mathematician and clergyman. Author of "Atheistical Objections Against the Being of God and His Attributes, Fairly Considered and Fully Refuted."

Runners-up List

The elegance and simplicity of Newtonian gravitation might be used as an argument for the existence of God. We could imagine universes with other gravitational laws and much more chaotic planetary interactions. But in many of those universes we would not have evolved—precisely because of the chaos. Such gravitational resonances do not prove the existence of God, but if he does exist, they show, in the words of Einstein, that while he may be subtle, he is not malicious.

—Carl Sagan, *Broca's Brain*

The following individuals received one vote each. Several may be more correctly classified as physicists or philosophers than as mathematicians: Ross Honsberger (who asserts his acceptance of Christianity on the cover of *Ingenuity in Mathematics*), Copernicus (a Catholic priest), Michael Faraday (experimental physicist and devout Sandemanian, a small Protestant group), W. de Sitter (also a priest), Lemaitre (20th-century cosmologist and Catholic abbe), Charles Misner, Raymond Smullyan, Wernher von Braun (a space scientist who said, "An outlook through this peephole at the vast mysteries of the universe should confirm your belief in the Creator."), Don Page, David Wagner (mathematician and confessional Lutheran), Jan Sengers, George Salmon (FRS, mathematician), Alan Sandage, George Berkeley (philosopher and Bishop), Søren Kierkegaard (philosopher and theologian), Immanuel Kant (philosopher, Pietist), Igor Shafarevich (mathematician), and Teilhard de Chardin (paleontologist, philosopher, and Jesuit priest).

Was Albert Einstein Religious?

Albert Einstein was mentioned by several respondents, although he is not considered religious in the traditional sense. Einstein himself wrote:

> It was, of course, a lie what you read about my religious convictions, a lie which is being systematically repeated. I do not believe in a personal God and I have never denied this but have expressed it clearly. If something is in me which can be called religious then it is the unbounded admiration for the structure of the world so far as our science can reveal it.
>
> I do not believe in immortality of the individual, and I consider ethics to be an exclusively human concern with no superhuman authority behind it. (*Albert Einstein: The Human Side*, edited by Helen Dukas and Banesh Hoffman, Princeton University Press, 1981)

On the other hand, there are many times in Einstein's life where he had some distinctly religious feelings. John Mitchell from San Diego, California, brought several of these quotations to my attention. The quotes are from *Ideas and Opinions by Albert Einstein* (Bonanza Books, New York, distributed by Crown Publishers, 1995).

After describing the historical development of religion from a "religion of fear" to a "religion of morality," Einstein notes:

> But there is a third stage of religious experience which belongs to all of them, even though it is rarely found in a pure form: I shall call it cosmic religious feeling. It is very difficult to elucidate this feeling to anyone who is entirely without it, especially as there is no anthropomorphic conception of God corresponding to it.... In my view, it is the most important function of art and science to awaken this feeling and keep it alive in those who are receptive to it.

Here are some additional related thoughts from Einstein:

> On the other hand, I maintain that the cosmic religious feeling is the strongest and noblest motive for scientific research.

> A contemporary has said, not unjustly, that in this materialistic age of ours the serious scientific workers are the only profoundly religious people. (*New York Times Magazine*, Nov. 9, 1930, pp. 1–4)

> [The scientist's] religious feeling takes the form of a rapturous amazement at the harmony of natural law, which reveals an intelligence of such superiority that, compared with it, all the systematic thinking and acting of human beings is an utterly insignificant reflection. This feeling is the guiding principle of his life and work.... It is beyond question closely akin to that which has possessed the religious geniuses of all ages. (*Mein Weltbild*, Amsterdam: Querido Verlag, 1934)

In a discussion of science and religion, Einstein says:

> But mere thinking cannot give us a sense of the ultimate and fundamental ends. To make clear these fundamental ends and valuations, and to set them fast in the emotional life of the individual, seems to me precisely the most important function which religion has to perform in the social life of man.
>
> The highest principles for our aspirations and judgments are given to us in the Jewish-Christian religious tradition.
>
> It is only to the individual that a soul is given. Science without religion is lame, religion without science is blind.
>
> And so it seems to me that science not only purifies the religious impulse of the dross of its anthropomorphism but also contributes to a religious spiritualization of our understanding of life. (From an address at Princeton Theological Seminary, May 19, 1939)

Here are some additional quotations from Einstein regarding religion:

> The interpretation of religion, as here advanced, implies a dependence of science on the religious attitude, a relation which, in our predominantly materialistic age, is only too easily overlooked. ("Religion and Science: Irreconcilable?" *The Christian Register,* June 1948)

> To be a Jew, after all, means first of all, to acknowledge and follow in practice those fundamentals in humaneness laid down in the Bible—fundamentals without which no sound and happy community of men can exist.

POSTSCRIPT **3**

Some Mind-Boggling Terminology

Many "History behind the Science Fiction" sections outline some of the more obscure and fascinating information available on the life of Pythagoras. To facilitate your reading, I define a list of theological terms used in these chapters and throughout the book. These terms are also frequently found in serious theomatic literature discussing the end of the universe, and I find the words interesting in their own right. I hope you find this list both amusing and enlightening.

1 *Aniconic*—referring to simple material symbols of a god, for example a pillar or block, not shaped into an actual image of human form.

2 *Anamnesis*—a remembrance, sometimes used in reference to past lives in reincarnation theologies. Pythagoras and Plato believed the soul pre-existed in past lives where it gained ideas useful in the present life.

3 *Antichthon*—A hypothetical second Earth on the opposite side of the Sun. Pythagoras and his followers asserted the existence of the Antichthon. See *counter-earth.*

4 *Apocrypha*—Writing of dubious authorship. A term specifically applied to those books in the Septuagint and Vulgate versions of the Old Testament which were not originally written in Hebrew and not counted genuine by the Jews. The Apocrypha were excluded from Sacred Canon by the Protestants because they have no well-grounded claim to inspired authorship.

5 *Aretalogy*—a story of miracles performed by a divine hero or god. Aretalogy often concerns the acts of a *thaumaturge.*

6 *Apocalypse*—the revelation of the future granted to St. John. Also the book of the New Testament in which St. John recorded this information. Apocalypse generally means "revelation," and it is sometimes applied to various Old Testament prophetic writing (particularly Ezekiel and Daniel), some *pseudepigraphic* writ-

ing (such as the books of Enoch), and some of the *Apocrypha* (such as four Esdras). Its subject is often *eschatological.*

7   *Armageddon*—the place of the last decisive battle at the Day of Judgment. (No relation to *Armagnac*, a superior brandy made in the district of France formerly called Armagnac.)

8   *Chthonic*—dwelling beneath the Earth.

9   *Catachthonian*—subterranean.

10   *Cenotaph*—a monument erected in honor of a deceased person whose body is elsewhere.

11   *Choical*—a Gnostic term for earthy.

12   *Counter-earth*—an opposite or secondary Earth in which Pythagoreans believed. See antichthon.

13   *Didache*—the name of a Christian treatise of the beginning of the 2nd century. Also the instructional element in early Christian theology, as distinct from kerygma or preaching.

14   *Doomsday*—the judgment day or end of the world.

15   *Dweomercraeft*—jugglery and magic art. Sorcery. Also see necromancy.

16   *Eisegesis*—the interpretation of Scripture by reading into it one's own ideas.

17   *Enatiodromia*—the process by which a belief becomes its opposite, and the subsequent interaction of the initial belief and its opposite. A good example of enatiodromia is seen in the psychology of Saul of Tarsus and his conversion to Christianity. Heraclitus believed that all beliefs eventually meet their opposites. In 1943, E. L. Mascall noted that one of the main tenets of Islam is submission, and by that enatiodromia, because Islam became "the most militant religion in history, for once the believer has made his submission, he sees himself as an instrument of the divine ruthlessness." An example of an enatiodromiacal reaction is the transition between the Middle Ages and the Renaissance. The repressive Middle Ages, with its increasing dogmatism and formalism, produced an enatiodromical reaction of the Renaissance.

18   *Epiclesis*—a part of the prayer of consecration where the Holy Spirit is invoked to bless the eucharist and/or communicants.

19   *Eschatologist*—one who studies *eschatology.*

20   *Eschatology*—theological science concerned with the four last things: death, judgment, heaven, and hell. The study of the end of history. Some Christians see the life of the individual Christian and the Church as a series of decisions with an eschatological nature.

21   *Eschaton*—the divinely ordained climax of history. In many of the *apocalyptic* cults, eschaton is always close at hand.

22   *Exegesis*—practical explanation or critical interpretation of Scripture.

23   *Feng-shui*—in Chinese mythology, a system of spirtual influences which inhabit the natural features of landscapes. Feng-shui is a kind of *geomancy* for selecting appropriate sites for houses and graves.

24   *Geomancy*—divination using signs derived from the earth. For example, the pattern produced by a pile of sand may be used to predict the future. Geomancy is

also divination by means of lines or figures formed by jotting down on a paper a number of dots at random. (See brief mention in Chapter 9.)

25   *Gigantology*—treatises about giants.

26   *Goety*—witchcraft or magic performed by the invocation and employment of evil spirits.

27   *Hermeneutic*—concerning interpretation, especially distinguished from *exegesis*. The study of the methodological principles of interpretation of the Bible.

28   *Hermetica*—works of revelation dealing with occult, theological, and philosophical subjects attributed to the Egyptian god Thoth (Greek Hermes Trismegistos, i.e., "Hermes the Thrice-Greatest"), who was said to be the inventor of writing and all the arts that depend on writing. (See Chapter 13.) During Hellenistic times, there was an increasing distrust of traditional Greek rationalism and the destruction of the line between science and religion. Hermes–Thoth was one of several gods to whom humans turned to for wisdom.

29   *Interimsethik*—the moral principles given by Jesus interpreted as a guide to humans expecting the imminent end of the world.

30   *Lullian*—belonging to the mystical philosophy of Raymon Lull (1234–1315). (See Chapter 9 for more on Lullists.)

31   *Kerygma*—preaching (see *Didache*) or proclamation of religious truth.

32   *Lungis*—*apocryphal* name of the Roman officer who pierced Jesus with a spear.

33   *Metempsychosis*—transmigration of human or animals souls into a new body (whether of the same or different species). A tenet of Pythagoreans and Buddhists.

34   *Meturgeman*—an interpreter of religious law. See also *Targum*.

35   *Monad*—the number one, historically used with reference to Pythagoras who regarded numbers as real entities and as primordial principles of existence. The term was adopted by Leibniz from Giordano Bruno (Chapter 13), with whom "monad" referred to material atoms and ultimate elements of psychical existence.

36   *Necromancy*—the art of communicating with the dead.

37   *Nobodaddy*—a disrespectful name for God, used by William Blake and others. For example, Joyce in *Ulysses* says, "Whether these be sins or virtues, old Nobodaddy will tell us at doomsday."

38   *Palingensia*—Pythagorean belief in rebirth after death, cyclical regeneration.

39   *Pareschatology*—theories about human life and physical death before the final resolution. In contrast, *eschatology* is the doctrine of the eschata or last things. In other words, pareschatology is the study of paraeschatata or next-to-last things, and therefore the human future between the current life and humans and their ultimate state. In a 1977 issue of *Theology Today* (XXXIV.182), there is a discussion of pareschatology as a doctrine of resurrection expanded to include "vertical" as opposed to "horizontal" reincarnation. In the 1977 *Times Literary Supplement* (April 1, 390/4), there is an examination of Western and Eastern parechatologies, and images of what happens between death and an ultimate state.

40 *Phyletism*—in the Orthodox Church, an excessive emphasis on the principle of nationalism in the organization of church affairs. Phyletism attaches greater importance to ethnic identity than to faith and worship.

41 *Preterist*—one who holds that the prophecies of the *Apocalypse* have already been fulfilled.

42 *Protology*—the study of origins. Some religions may place more stress on protology then *eschatology*.

43 *Pseudepigrapha*—books bearing a false title or ascribed to the incorrect author. The term is specifically applied to certain Jewish writings dating to the beginning of the Christian era but ascribed to various patriarchs and prophets of the Old Testament.

44 *Scatophagy*—the religious practice of eating excrement.

45 *Sciomancy*—divination by communicating with the shades of the dead.

46 *Targum*—Aramaic translations of parts of the Old Testament, made after the Babylonian captivity, at first preserved by word-of-mouth, and committed to writing starting in A.D. 100.

47 *Thaumaturge*—a worker of miracles and wondrous things. Some historians describe Pythagoras as a thaumaturge.

48 *Theomatic*—a melding of theology and mathematics. Similarly, the word *theomata* is defined as works arising from this melding of theology and math.

49 *Theurgy*—a system of magic to facilitate communication with beneficent spirits and produce miraculous effects. The art or science of compelling or persuading God to do something. Theurgy is sometimes known as "white magic," while *goety* is "black magic." Theurgy is also defined as the operation of a divine or supernatural agency in human affairs.

Author's Musings

These Notes as I see them, relate not to lectures but to feeling. I'm sure my readers differ from me on many things, but I hope that we share the essence of wonder and longing for what we may never quite understand.
　　　　　　　　　　　　—Piers Anthony, 1991, *Virtual Mode*

Piers Anthony, one of science fiction and fantasy's most prolific talented writers, established the interesting practice of placing an "Author's Note" section at the end of some of his books. In these notes, he gives that slice of his life occurring during the writing of a current novel, complete with discussions of social issues and unfinished thoughts. In keeping with Piers (with whom I collaborated on a recent science fiction novel), I started this practice in my book *Chaos and Wonderland* and continue it here by providing you a slice of some of the mail I have received while writing *The Loom of God*. Also included are miscellaneous recent thoughts, and reader responses to subjects in my previous books.

A Double Whammy?

> *If one was able to replay the whole evolution of animals, starting at the Cambrian and, to satisfy Laplace, move one animal two feet to its left, there is no likelihood that the result would be the same. There might be no conquest of land, no emergence of mammals, certainly no human beings.*
> 　　　　　　　　　　　　—Evolutionist John Maynard Smith

In Chapters 10 and 21, I discuss theories relating the demise of the dinosaurs to an impact of a huge comet or asteroid. The January 9, 1995, issue of *Time* presents a recent view on this subject. One new theory holds that clouds of sulfuric acid, formed from the debris thrown into the atmosphere by the impact, were the primary cause of extinctions. Another theory suggests that impact also caused extreme volcanic activity on the opposite side of the Earth, creating a double whammy that killed the dinosaurs. Jon Hagstrum of the U.S. Geological Survey says, "This would be the best way to trigger worldwide mass extinctions because you have both Earth hemispheres affected."

The likely site of the impact is centered below the town of Chicxulub on the northern tip of Mexico's Yucatan peninsula. The Yucatan rock around Chicxulub contains large quantities of sulfur. The impact 65 million years ago would have vaporized the sulfur and spewed 100 billion tons of it into the air where it mixed with moisture to form drops of sulfuric acid. These drops reflected sunlight back into space to drop temperatures to near freezing for decades. Kevin Baines of the Jet Propulsion Laboratory comments, "If the asteroid had struck almost any other place on Earth, it wouldn't have generated this tremendous amount of sulfur. Dinosaurs would still be roaming the Earth."

How could debris from the impact have traveled high enough into the stratosphere to block out the sun or destroy the ozone layer? Computer simulations suggest that a large asteroid or comet slamming into the ocean floor is capable of producing produce a "hypercane"—the mother of all hurricanes—20 miles high and with winds approaching 500 miles an hour. This would have been sufficiently strong to spread debris all over the Earth. For more information, see: Svitil, K. (1995) Hurricane from Hell. *Discover* April 16(4): 26.

Death Star

> *The period of history which is commonly called "modern" has a mental outlook which differs from that of the medieval period in many ways. Of these, two are the most important: the diminishing authority of the Church, and the increasing authority of science.*
>
> —Bertrand Russell

In Chapter 10, I mentioned that a Death Star companion to our Sun was once hypothesized as the cause for recurring extinctions in the history of life. Another kind of Death Star is described in the January 20, 1995, issue of *Science* 267(5196):334.

Could a nearby supernova have caused the mass extinction on Earth 251 million years ago—way before the dinosaurs? Researchers suggest a supernova exploding within 30 light-years of Earth is expected to occur once every 240 million years. Even at this distance, the exploding star could bombard the Earth with enough radiation to destroy the ozone layer, thereby exposing photosynthetic creatures at the low end of the food chain to lethal doses of ultraviolet radiation. Currently, scientists debate whether the biological effects of the ozone loss would be sufficient to kill 90% of Earth's species—as happened at the end of the Permian period 251 million years ago.

Averting Asteroid Doom

In May 1995, a group of astronomers, physicists, and engineers met at Lawrence Livermore National Laboratory to discuss the risks of asteroids and comet crashes. Many methods for protecting the Earth were discussed including non-nuclear schemes for breaking asteroids into little pieces or for forcing them off course. Other "spaceguard" suggestions included: automated charge-coupled devices mounted on telescopes for precise asteroid detection, the use of high-velocity projectiles strung together with strong fibers ("kinetic energy cookie cutters"), the use of rockets to crash into asteroids, the use of solar collectors for heating asteroids' surfaces, and various methods for shoving small asteroids into larger ones.

One of the scariest predictions advanced at the meeting was the Earth should experience a major asteroid-triggered tsunami every 4000 years. For more information, see: Hill, D. (1995) Gathering airs schemes for averting asteroid doom. *Science*. June 268(5217): 1562.

Marriage in Cyberspace

Robert Schloss of Yorktown Heights, New York, has brought several interesting facts to my attention regarding virtual reality. In my book *Mazes for the Mind*, I discuss computers allowing us to enter lifelike, computer-generated realities. Strap on a pair of goggles and you gaze into a three-dimensional world limited only by the speed with which your computer can change the images in response to your eye and head motions. Some call this sense of actually being present within a new reality a "virtual reality" or cyberspace.

On August 20, 1994, 25-year-old Monika G. Liston and 33-year-old Hugh H. Jo were married in the world's first virtual reality wedding. The couple and their minister donned computerized headgear in pods at CyberMind Virtual Reality Center in San Francisco. They immediately became immersed in a recreation of the lost city of Atlantis containing a palace, chariots, carousels, and doves. The bride, a CyberMind executive assistant, told BusinessWeek: "We've been married in our ultimate dream palace."

Two other interesting factoids. In the 1980s, a new religious denomination was formed every week (source: *Newsweek*), 52% of 3000 11-year-olds interviewed in New York state would prefer to be without their father rather than without their televisions (source: Minty, L. (1992) *Television Viewing and Your Brain*)

Chaotic Dreams

Michael Seargant writes to me:

> I was captivated by the "Latööcarfian Dream" images in your book *Chaos in Wonderland: Visual Adventures in a Fractal World* (St. Martin's Press, NY). I therefore was compelled to write a program for IBM PCs to reproduce the images.

Mike says that the free programs are available by anonymous ftp:spanky.triumf.ca [pub.fractals.programs.ibmpc]. By WWW, the url is: http://spanky.triumf.ca/pub/fractals/programs/ibmpc/ganymede.zip. For additional information, contact: Mike Sargent, University of Vermont Student Health Center, 425 Pearl St., Burlington, VT, 05401 or e-mail him at msargent@moose.uvm.edu.

A Universe Called JUMBLE

In this book's Introduction, I wrote: "Imagine a universe called JUMBLE where Kepler looks up into the heavens and finds that most planetary orbits cannot be approximated by ellipses ..." You should know that although most orbiting bodies follow elliptical orbits, a few confound scientists with their strangeness. For example, Miranda, a moon of Uranus, once spent millions of years in a chaotic orbit, tumbling erratically like a trapeze artist high on LSD. Hyperion, a moon of Saturn, is now tumbling chaotically, and chaotic orbits explain how asteroids reach Earth in the form of meteors. Many other satellites of the outer planets may have orbited chaotically before ending up in their present, predictable orbits.

Mathematical Cranks and the Geometry of Heaven

Some of you will be interested in Underwood Dudley's *Mathematical Cranks* (1992, Mathematical Association of America: Washington, D.C.). In it are the musings of slightly mad mathematicians. One favorite formula relates the value of π to various physical constants such as the speed of light (c) and Planck's constant (h):

$$\pi = \left(\frac{E}{1/2mc^2}\right)^{1/2} [J_\lambda \cdot \lambda^5 (e^{hc/k_\lambda T} - 1)]$$

Another relates the golden ratio ϕ with π in a compact formula:

$$\frac{6}{5}\phi^2 = \pi$$

Is there anything wrong with this formula?

My favorite chapter is on *Matrix Prayers* designed by a priest of the Church of England. The priest regularly prayed to God in mathematical terms using matrices, and he taught the children in his church to pray and think of God in matrices. The book goes into great detail regarding "revelation matrices," "Polite Request Operators," and the like. The priest finally derives a beautiful prayer which he succinctly writes as:

$$P < R\{S\} \rightarrow \{U\}, \text{ all } r > 0 >$$

He says, "This prayer should be sufficiently concise to be acceptable to Christ, yet every single Christian inhabitant of Northern Ireland has been separately included." The chapter concludes with information regarding the geometry of heaven.

God and the Physical Constants

The null set is also a set; the absence of a god is also a god.
—A. Moreira

Many physicists have been excited about the possibility of establishing units for time, mass, and length which are independent of specific bodies and which would have a meaning for all time and civilizations. The units would even have a meaning for extraterrestrial civilizations. By combining the velocity of light c with the gravitational constant G and Planck's constant h, physicists compute:

Planck length = $(Gh/c^3)^{1/2} = 10^{-33}$ cm
Planck time = $(G/c^5)^{1/2} = 10^{-43}$ s
Planck mass = $(hc/G)^{1/2} = 10^{-5}$ g

Notice that the values have enormously different size scales, and this disparity has intrigued physicists for a century. Physicists have also wondered if the fundamental constants of nature—such as the mass of the electron, velocity of light, charge of an electron, gravitational constant, and Planck's constant—are connected in some subtle way. I believe the current thinking is that the constants are independent entities. (Source: Klotz, I. (1995) Number mysticism in scientific thinking. *Mathematical Intelligencer* 17(1):43–51)

One unusual paper in a prestigious chemical journal reported that many products and quotients of fundamental constants have values very close to the number 3:

Function	Numerical Value
c	2.9976×10^{10}
$(\epsilon/m_0)^{1/2}$	2.9995×10^1
$m_0/(2\pi\epsilon)$	3.0009×10^9
$\sqrt{m_0}$	2.9990×10^{-14}
$2\pi\epsilon$	2.9971×10^{-9}
$3Gc/2$	2.9967×10^3

Author J. E. Mills claims that the high occurrence of values extremely close to the number 3 is "amazing." (Do you think it is amazing?) In the table, m_0 is the mass of an electron, and ϵ the electronic charge. (Source: Mills, J. E. (1932) Relations between fundamental physical constants. *J. Physical Chemistry* 36:1089–1107.)

Vampire Numbers

If we are to believe best-selling novelist Anne Rice, vampires resemble humans in many respects, but live secret lives hidden among the rest of us mortals. In *Keys to Infinity* I asked readers to consider a numerical metaphor for vampires. I call numbers like 2187 "vampire numbers" because they're formed when two progenitor numbers 27 and 81 are multiplied together ($27 \times 81 = 2187$). Note that the vampire, 2187, contains the same digits as both parents, except that these digits are subtly hidden, scrambled in some fashion. Similarly, 1435 is a vampire number because it contains the digits of its progenitors, 35 and 41, since ($35 \times 41 = 1435$). These vampire numbers secretly inhabit our number system, but most have been undetected so far. John Childs wrote to me of a 40-digit vampire number he discovered using a Pascal program on a 486 personal computer. His amazing vampire number is 98765432198765432198 × 98765432198830604534 = 9754610597415368368844499268390128385732.

THE LOOM
OF GOD

I conclude these "Author's Musings" with "Satan is a Mathematician," a poem from
Keith Allen Daniels.

The tattoo demon laughed.
"I shall inscribe you with pi—
a pi whose digits are fractal glyphs
of transcendental agony, whose serifs
are influorescent with infinities.
And I shall render it with all
the panache of a pointillist
creating continua from the discrete.
But where to begin? The umbilicus
or the anus? The alpha or the omega?"

"Hey, wait a minute!" cried
the mathematician, and the demon
raised an eyebrow. "Pi's an irrational number
with a nonrepeating decimal.
Such a task would take an eternity!"

"Imagine that," said the demon,
and smiling smugly, it poised a talon
tapering to a single atom, plucked lint
from the navel of its flinching victim.

Playing his last card, the mathematician
rose up on his elbow. "Have you really
thought about this? When the flesh
of one man emblazoned subsumes the infinite,
you will have modeled God from numbers
and I will destroy you!"

The other eyebrow twitched. "Well, then,
I shall adorn you with the closest
rational approximation of pi."

"Shit!" said the mathematician.

"As you wish," replied the demon,
and began with the anus.

Smorgasbord for Computer Junkies

COMPUTE THE FIRST 100 TRIANGULAR NUMBERS (CHAPTER 1)

```
10 REM Compute Triangular Numbers
20 FOR I = 1 TO 100
30     T = (1/2)*I*(I + 1)
40     PRINT I, T
50 NEXT I
60 END
```

COMPUTE UMBUGIO'S "END-OF-THE-WORLD" NUMBERS (CHAPTER 2)

```
10 REM Compute Umbugio's "End-of-the-World" Numbers
20 FOR N = 1 TO 10
30     A = 1492¬N − 1770¬N − 1863¬N + 2141¬N
40     PRINT N, A
50 NEXT N
60 END
```

Note: Can you write a program to check the divisibility of Umbugio's numbers? Calculating the powers, as in this short program, will quickly cause numerical overflows. Can you design a program to avoid this problem?

COMPUTE POLYGONAL NUMBERS, TRIANGULAR TO HEXAGONAL (CHAPTER 3)

```
10 REM Compute Polygonal Numbers, Triangular to Hexagonal
20 FOR R = 3 TO 6
30     FOR N = 1 TO 6
40         P = (N/2) * ((N − 1)*R − 2*(N − 2))
50         IF R = 3 THEN PRINT "Triangular:" P
60         IF R = 4 THEN PRINT "Square:" P
70         IF R = 5 THEN PRINT "Pentagonal:" P
80         IF R = 6 THEN PRINT ".Hexagonal:" P
90     NEXT N
100 NEXT R
110 END
```

Note: Joseph F. Pycior from San Jose, California, writes, "I converted the polygonal number program to Fortran, and found it ironic that I had to use floating point numbers to produce the polygonal numbers, because integers were inadequate. Pythagoras wouldn't have approved. I computed as high as 11-sided polygons and was amazed the numbers didn't become large as fast as I expected."

COMPUTE VON FOERSTER DOOMSDAY CURVE (CHAPTER 4)

```
10 REM Compute Doomsday Date
20 FOR T = 0 TO 2000 STEP 100
25 REM FOR T = 1980 TO 2000
30     N = (1.79E11)/((2026.87 − T)¬.99)
40     PRINT T, N
50 NEXT T
60 END
```

COMPUTE UNITED STATES POPULATION VIA PEARL-REED (CHAPTER 4)

```
10 REM Compute United States Population via Pearl-Reed
20 FOR T = 1790 TO 1910
30     X = 210 /(1 + 51.5*EXP(−0.0313*(T − 1790)))
40     PRINT T, X
50 NEXT T
60 END
```

```
10 REM Generate Pythagorean Triangles
15 PRINT "Pythagorean Triples (side lengths):"
20 FOR U = 1 TO 6
30     FOR V = U + 1 TO 7
40         X = 2*U*V
50         Y = ABS(U¬2 − V¬2)
60         Z = V¬2 + U¬2
70         PRINT X,Y,Z
80     NEXT V
90   NEXT U
100 END
```

COMPUTE PRAYING TRIANGLES (CHAPTER 5)

```
10 REM Compute Lengths for Triangles that Pray
15 REM Print First Four Examples
17 REM High precision needed for larger examples
20 D = (SQR(2) + 1)*(SQR(2) + 1)
30 S = 1
40 FOR I = 1 TO 4
50     S = S*D
60     S = INT(S)
70     A = SQR((S*S)/2)
80     A = INT(A)
90     B = A + 1
100    C = SQR(A*A + B*B)
110    PRINT "PYTHAGOREAN TRIANGLE (A,B,C):",A,B,C
120 NEXT I
130 END
```

FIND ALL ST. AUGUSTINE NUMBERS (CHAPTER 6)

```
5  REM Search for St. Augustine Numbers
10 REM (Cubical Narcissistic Numbers)
15 REM Search all 3-digit integers
20 FOR I = 100 TO 999
30     A = INT(I/100)
40     B = INT(I/10) − 10*A
```

```
50    C = I − 100*A − 10*B
55    IF I <> A**3 + B**3 + C**3 THEN 90
60    PRINT "Narcissistic Number"; I
70    PRINT "Equals";A**3;"+";B**3;"+";C**3
80    PRINT
90  NEXT I
100 END
```

FIND PERFECT NUMBERS (CHAPTER 7)

```
10 REM Compute Perfect Numbers
15 FOR N = 2 TO 100
20    S = 0
30    FOR D = 1 TO N/2
40       IF INT(N/D) <> N/D THEN 60
50       S = S + D
60    NEXT D
70    IF S <> N THEN 90
80    PRINT N; "is a perfect number"
90 NEXT N
100 END
```

FIND AMICABLE NUMBERS (CHAPTER 7)

```
100 REM Find Amicable Numbers
110 FOR L = 1 TO 8000
120    S = 0
130    FOR D = 1 TO L/2
140       IF L/D <> INT(L/D) THEN 160
150       S = S + D
160    NEXT D
170    IF S <= L THEN 260
180    B = S
190    T = 0
200    FOR F = 1 TO B/2
210       IF B/F <> INT(B/F) THEN 230
220       T = T + F
230    NEXT F
240    IF T <> L THEN 260
```

```
260 NEXT L
270 END
```

SOLVE TURK AND CHRISTIAN PROBLEM (CHAPTER 8)

```
10 REM Solve Turk and Christian Problem
20 REM A "1" in the array means the person has died.
30 REM Each row of output indicates the cumulative deaths.
40 DIM A(30)
50 J = 0
60 FOR I = 1 TO 15
70     J = J + 13
80     IF J > 30 THEN J = J - 30
90     A(J) = 1
100    FOR K = 1 TO 30
110        PRINT A(K);
120    NEXT K
130    PRINT
140 NEXT I
150 END
```

COMPUTER DOOMSDAY ORBITS OF ASTEROIDS (CHAPTER 10)

```c
/* Compute Doomsday Orbits of Asteroids */
/* Simulation: Is the GPPA necessary? */
#include <math.h>
#include <stdio.h>
main ()
{
    int i;
    float a, cx, cy, theta, x, y;
  /* Draw Earth Orbit at Center */
  for (theta=0; theta <= 6.3; theta=theta+.1) {
    x = .5*cos(theta);
    y = .5*sin(theta);
    printf("%f %f\n",x,y);
  }
  /* Position Earth on its Orbital Drawing */
```

```c
x = .5*cos(0);
y = .5*sin(0);
printf("%f %f\n",x,y);
/* Draw Random Elliptical Asteroid Orbits Near Earth */
for (i = 0; i <= 10; i++)
 /* Generate random number to control eccentricity of orbit */
 a = (float)rand()/32767.;
 /* Draw Asteroid Orbit */
  for (theta = 0; theta <= 6.3; theta = theta + .1) {
    x = cos(theta);
    y = a*sin(theta);
    printf("%f %f\n",x,y);
}
 /* Randomly Position Asteroids on their Elliptical Orbits */
 theta = 6.28*(float)rand()/32767.;
 x = cos(theta);
 y = a*sin(theta);
 printf("%f %f\n",x,y);
```

FIND PRIME FACTORS OF URANTIA NUMBERS (CHAPTER 12)

```basic
10 REM Find Prime Factors of Urantia Numbers
20 REM Urantia Number to be factored:
30 A = 611121
35 REM Also try A = 5342482337666
40 IF ABS(A) <= 1 THEN 210
50 N = INT(ABS(A))
60 REM Find the prime factors and print
70 B = 0
80 FOR I = 2 TO N/2
90    IF N/I > INT(N/I) THEN 170
100   B = B + 1
110   IF B > 1 THEN 130
120   PRINT "Prime Factors of";N; "are:"
130   PRINT I
140   N = N/I
150   IF N = 1 THEN 210
160   I = I - 1
170 NEXT I
```

```
180 IF N <> INT(A) THEN 120
190 PRINT N; "is a prime number."
210 END
```

FRACTAL GENERATOR (CHAPTER 13)

```
100 REM Julia set program
110 CR = 0        'Real part of C
120 CI = 1        'Imaginary part of C
130 X1 = −1.5     'Boundaries of plot
140 X2 = 1.5
150 Y1 = −1.5
160 Y2 = 1.5
170 SCREEN 12     'Assume VGA color graphics mode
180 W% = 640      'Screen width
190 H% = 480      'Screen height
200 KMAX% = 64    'Bailout condition
210 FOR I% = 0 TO W% − 1
220     FOR J% = 0 TO H% − 1
230         K% = 0
240         C% = KMAX%
250         X = X1 + I% * (X2 − X1) / W%
260         Y = Y2 − J% * (Y2 − Y1) / H%
270         XS = X * X
280         YS = Y * Y
290         WHILE K% < KMAX%
300             Y = Y + Y
310             Y = X * Y + CI
320             X = XS − YS + CR
330             XS = X * X
340             YS = Y * Y
350             K% = K% + 1
360             IF XS + YS > 4 THEN C% = K%: K% = KMAX%
370         WEND
380         PSET (I%, J%), C% MOD 16
390     NEXT J%
400 NEXT I%
410 END
```

I personally enjoy programming a computer using the language C these days, but I know most hobbyists use BASIC. Therefore, I've included a BASIC program that should allow you to compute and display Julia sets. Computer-literate readers should be able to further modify it for their own use. The program has been tested with GW-BASIC, QuickBASIC, VisualBASIC, and PowerBASIC on the IBM PC platform and with QuickBASIC on the Macintosh. Users may wish to adjust the screen mode and size in lines 170-190 for the various platforms. The program produces a Julia dendrite in about 12 seconds using PowerBASIC 3.0 on a 486-DX2/66 machine. The values for CR and CI may be adjusted to produce different Julia sets. Professor J. Clint Sprott, well-known for his work in strange attractors, wrote the preceding code.

COMPUTE QUIPU PATTERN (CHAPTER 14)

```
/* Simulate Quipu Structure */
#include <math.h>
#include <stdio.h>
main()
{
    int i;
    float a,cx,cy,theta,x,y;

    /* Draw Main Cord of Quipu */
    for (theta=3.14; theta <= 7.28; theta = theta + .1) {
    x = cos(theta);
    y = sin(theta) + 1;
    printf("%f %f\n",x,y);
}

    /* Position Pendant Cords */
    for (theta = 3.14; theta <= 6.28; theta = theta + .1) {
    x = cos(theta);
    y = sin(theta) + 1;
    printf("%f %fn",x,y);
    x = 2*cos(theta);
    y = 2*sin(theta) + 1;
    printf("%f %f\n",x,y);

    /* subsidiary cords */
    x = 1.5*cos(theta);
    y = 1.5*sin(theta) + 1;
    printf("%f %f\n",x,y);
    x = 2.2*cos(theta + .05);
```

```
y = 2.2*sin(theta + .05) + 1;
printf("%f %f\n",x,y);
    }
    }
```

COMPUTE FIBONACCI NUMBERS (CHAPTER 15)

```
10 REM Compute Fibonacci Numbers
20 DIM F(40)
30 F(1) = 1
40 F(2) = 1
50 FOR N = 1 TO 38
60    F(N + 2) = F(N + 1) + F(N)
70 NEXT N
80 FOR N = 1 TO 40
90    PRINT F(N)
100 NEXT N
110 END
```

COMPUTE FIBONACCI SPIRALS (CHAPTER 15)

```
10 REM Compute Fibonacci Spirals
20 G = (1 + SQR(5))/2
30 T = 2*3.1415926/G
35 REM Experiment with different values of K
40 K = 2
50 R2 = K*G/2
55 REM 2000 Circles
60 FOR I = 1 TO 2000
70    R = K * SQR(I)
80    A = T*I
85    REM C1 and C2 are circles' centers
90    C1 = R*SIN(A)
100   C2 = R*COS(A)
150   REM DRAW A CIRCLE, T2 in degrees
200   FOR T2 = 1 TO 370 STEP 10
300     X = R2*COS(T2) + C1
400     Y = R2*SIN(T2) + C2
```

```
500    PRINT X, Y
600    NEXT T2
700 NEXT I
900 END
```

COMPUTE NUMERICAL GARGOYLES (CHAPTER 17)

C Program Code

```c
/* Create numerical gargoyles. Magnify "creatures" as needed. */
#include <stdio.h>
#include <math.h>
main()
{
    float r; /* random number, 0-1 */
    short c[513][513]; /* Arrays for holding 1 and 0 values */
    short ch[513][513];
    int size, time_steps, steps, i,j,k,sum;
    size = 80; /* Use larger sizes for nicer images */
    /* Controls how many steps gargoyle is to evolve */
    time_steps = 20;
    /* Initially seed space with random 0's and 1's */
    for(i = 0; i <= size; i++)
     for(j = 0; j <= size; j++) {
        r = (float) rand()/32767.;
        if (r >= .5) c[i][j] = 0; if (r <= .5) c[i][j] = 1;
     }
    /* perform simulation based on twisted majority rules
       to form gargoyle objects in 2-D */
    for(steps = 1; steps < time_steps; steps++) {
        for(i = 1; i < size; i++) {
        for(j = 1; j < size; j++) {
         /* compute sum of neighbor cells */
            sum = c[i + 1][j + 1] +
                c[i - 1][j - 1] +
                c[i - 1][j + 1] +
                c[i + 1][j - 1] +
                c[i + 1][j] + c[i - 1][j] +
                c[i][j + 1] + c[i][j - 1] +
                c[i][j];
```

```c
            if (sum == 9) ch[i][j] = 1;
            if (sum == 8) ch[i][j] = 1;
            if (sum == 7) ch[i][j] = 1;
            if (sum == 6) ch[i][j] = 1;
        /* Notice "twist" in rules which destabilizes
           creature boundaries. */
        if (sum == 5) ch[i][j] = 0;
        if (sum == 4) ch[i][j] = 1;
        if (sum == 3) ch[i][j] = 0;
        if (sum == 2) ch[i][j] = 0;
        if (sum == 1) ch[i][j] = 0;
        if (sum == 0) ch[i][j] = 0;
        }
   }
   for(i = 0; i <= size; i++) for(j = 0;j <= size; j++) c[i][j]
    = ch[i][j];
   }
   /* If you like, you can make a movie of all frames */
   /* to show gargoyle evolving.                      */
   printf("A plot of last frame of simulation\n");
   for(i = 0; i <= size; i++) {
      for(j = 0; j <= size; j++) {
          /* Crude attempt to draw creatures using characters */
          /* Better to use hi-res graphics package        */
          if (c[i][j] == 1) printf("*") ; else printf(" ");
      }
      printf("\n");
   }
}
```

BASIC Program Code

```basic
10 REM Create numerical gargoyles
15 REM Magnify "creatures" as needed
20 REM R is a random number, 0-1
30 REM C and C1 - Arrays for holding 1 and 0 values
35 DIM C(81,81)
40 DIM H(81,81)
50 REM Use larger size, S, for nice images
60 S = 80
```

毀

```
70 REM T Controls how many steps gargoyle is to evolve
80 T = 20
90 REM Initially seed space with random 0's and 1's
100 FOR I = 1 TO S
120     FOR J = 1 TO S
130         R = RND
140         IF R >= .5 THEN C(I,J) = 0
150         IF R <= .5 THEN C(I,J) = 1
160     NEXT J
170 NEXT I
175 REM Perform simulation based on twisted majority rules
180 REM to form gargoyle objects in 2-D
190 FOR K = 1 TO T
200     FOR I = 2 TO S −1
210         FOR J = 2 TO S − 1
220             REM compute sum of neighbor cells
230             A = C(I+1,J+1) + C(I−1,J−1) + C(I−1,J+1)
240             A = A + C(I+1,J−1) + C(I+1,J) + C(I−1,J) + C(I,J+1)
250             A = A+C(I,J−1) + C(I,J)
260             IF A = 9 THEN H(I,J) = 1
270             IF A = 8 THEN H(I,J) = 1
280             IF A = 7 THEN H(I,J) = 1
290             IF A = 6 THEN H(I,J) = 1
295             REM Notice "twist" in rules which destabilizes
297             REM blob boundaries.
300             IF A = 5 THEN H(I,J) = 0
310             IF A = 4 THEN H(I,J) = 1
320             IF A = 3 THEN H(I,J) = 0
330             IF A = 2 THEN H(I,J) = 0
340             IF A = 1 THEN H(I,J) = 0
350             IF A = 0 THEN H(I,J) = 0
360         NEXT J
370     NEXT I
380     REM Swap values in arrays
390     FOR I = 1 TO S
400         FOR J = 1 TO S
410             C(I,J) = H(I,J)
415         NEXT J
```

420 NEXT K

430 REM If you like, you can make a movie of all frames

440 REM to show gargoyles evolving.

450 PRINT "To plot the gargoyle, place a dot wherever the"

460 PRINT "C array has a value of 1 in it as you scan"

470 PRINT "I and J from 0 to S."

480 END

SIMULATE SIZE DISTRIBUTION OF MOON CRATERS (CHAPTER 21)

```c
/* Simulate Size Distribution of Moon Craters */
#include <math.h>
#include <stdio.h>

main()
{
    int i;
    float r,cx,cy,theta,x,y;
    for(i = 0; i < 80; i++) {
     /* Randomly determine center of each crater */
     cx = (float)rand()/32767.;
     cy = (float)rand()/32767.;
     /* Distribution of sizes is 1/r**2 via
        the Czarnecki transformation */
     r = (float)rand()/32767.;
     r = 0.01*tan(r*3.1415/2.);
     /* Draw Crater */
     for (theta = 0; theta <= 6.3; theta = theta + .1) {
       x = r*cos(theta) + cx;
       y = r*sin(theta) + cy;
       /* Data points trace out circular craters */
       printf("%f %f\n",x,y);
     }
    }
}
```

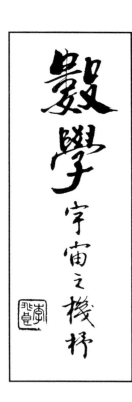

Notes

There's a world behind the world we see that is the same
world but more open, more transparent, without blocks ...
To touch this world no matter how briefly is a help in life.
　　　　　—Gary Snyder, *The Practice of the Wild*

At all the moments of death, one lives over again his past life with a rapidity
inconceivable to others. This remembered life must also have a last moment,
and this last moment its own last moment, and so on, and hence, dying is
itself eternity, and hence, in accordance with the theory of limits, one may
approach death but can never reach it.
　　　　　—Arthur Schnitzler, *Flight Into Darkness*

INTRODUCTION

1. Mathematician Stanislaw Ulam estimated that mathematicians publish 200,000 theorems every year, and I have increased this figure to reflect the growing number of mathematical publications. The majority of these theorems are probably ignored, and only a minuscule fraction comes to be understood and believed by any sizable group of mathematicians.
2. For those of you interested in learning more about the zany adventures of Mr. Plex, see *Black Holes, A Traveler's Guide* [Pickover, C. (1996), John Wiley and Sons, New York].

CHAPTER 1

1. In this chapter, you sip on Kylonian brandy. The Kylonian people really did exist during the time of Pythagoras and played a role in the life of Pythagoras (see Chapter 3). However, don't try to search for this word in your dictionary. The closest entry you'll find is "Kyloe," a small, shaggy breed of cattle with long horns raised in the Highlands and Western Islands of Scotland.

　　In order to better understand why the Kylonians are not mentioned in the best English dictionaries but rather in several history books, I spoke with Greek scholar G. Barry Grisdale. In

Celt and Greek, "ky" and "cy" are sometimes interchangeable. "Cylonians" were the followers of the mighty Cylon of Croton who, in 490 B.C., chased Pythagoras out of Croton to Metapontum where he later died.

2. The Appendix contains a BASIC program listing for computing triangular numbers.

CHAPTER 3

1. There are two kinds of hexagonal numbers often discussed in the literature. Centered hexagonal numbers created by nesting "concentric" hexagons can be generated using: $H_c(n) = 3n(n - 1) + 1$, $n = 1,2,3,\ldots$. Standard hexagonal numbers, as illustrated on p. 47, are generated using $H(n) = n(2n - 1)$, $n = 1,2,3,\ldots$. For more on hexagonal numbers, see Pickover, C. (1991, *Computers and the Imagination*. St. Martin's Press: New York.

2. For information on other polygonal numbers, and extension to higher dimensions such as pyramids and 4-dimensional figures, consult the "References" section. The Appendix contains a BASIC program listing for computing various polygonal numbers.

CHAPTER 4

1. Using these kinds of equations, after a few thousand years would humanity form a solid ball expanding away from Earth at the speed of light?

CHAPTER 5

1. Numbers in base 2 are called binary numbers. The digits used are 0 and 1. Each digit of a binary number represents a power of 2. The rightmost digit is the 1's digit, the next digit to the left is the 2's digit, and so on. In other words, the presence of a "1" in a digit position of a number base 2 indicates that a corresponding power of 2 is used to determine the value of the binary number. A "0" in the number indicates that a corresponding power of 2 is absent from the binary number. For example, the binary number 1111 represents $(1 \times 2^3) + (1 \times 2^2) + (1 \times 2^1) + (1 \times 2^0) = 15$. The binary number 1000 represents $1 \times 2^3 = 8$. Here are the first eight numbers represented in binary notation: 0000, 0001, 0010, 0011, 0100, 0101, 0110, 0111,...

2. In pre-Columbian South America, the Maya also believed in a seven-layered sky. In Egypt, there were seven paths to heaven. Osiris leads his father through seven halls of the netherworld.

CHAPTER 7

1. Plato's relationships are in stark contrast with Pythagoras' marriage. Plato was a philosopher (427 B.C.) who used many of Pythagoras' ideas and who wrote on political, ethical, metaphysical and theological questions. Plato, one of the great fathers of Western thought, did not have a good marriage or significant respect for women. George Sarton in *A History of Science* says,

Plato was not necessarily a pederast in the physical sense, but he was almost certainly homosexual. He never married, and though he speaks occasionally of sexual relations between men and women, he does so without emotion. His tender feelings were reserved for homosexual relations. He was somewhat of a woman hater. That is revealed many times in his writings.

2. Albertus Magnus was often called *doctor universalis* for the universality of his knowledge. He was canonized as a saint in 1931. He should not be confused with O. Magnus (1469), a Catholic priest who produced the first detailed map of Scandinavia.

CHAPTER 8

1. The *Turks and Christians* puzzle you described to Theano and Mr. Plex is in the literature and described in Mott-Smith's book listed in the "References" section. (The Appendix of *The Loom of God* contains a BASIC program listing for solving the puzzle.) In the god/mortal decimation problem Theano solved, the solution is 14 for the gods, and 13 for the mortals.

   In this chapter, you described the Tarantula Nebula as the home of the transfinites. The Tarantula Nebula is a complex of gas and dust in the Large Magellanic Cloud, one of the Milky Way's two companion galaxies. As you told Mr. Plex and Theano, although the Tarantula Nebula is 190,000 light-years away, it is visible to the naked eye for southern hemisphere viewers. The nebula is some 800 light-years across and contains enough matter to create 500,000 stars. A cluster of 100 supergiant stars is located at the center of this gargantuan nebula. Interestingly, if the Tarantula Nebula were as close as the Orion Nebula, it would cover an area of the sky 60 times that of the moon.

CHAPTER 9

1. Kurzweil, in *The Age of Thinking Machines*, comments about RACTER's book: "RACTER's prose has its charm, but is somewhat demented-sounding, due primarily to its rather limited understanding of what it is talking about."

CHAPTER 11

1. *Adipocere*—the gruesome, soapy substance to which a corpse buried in moist ground converts after many days. A greyish white fatty substance, chiefly Margarate of Ammonia, spontaneously generated in dead bodies buried in moist places, supposed to be produced by the reaction of ammonia upon margarine, oleine of the animal fat, and muscular fiber.

CHAPTER 12

1. For more information, see the "References" section. The Urantia Foundation can be contacted at 533 Diversey Parkway, Chicago, Illinois 60614.

CHAPTER 15

1. See Barret and McKay (1987) for more information on Fibonacci patterns. (The Appendix contains a BASIC program listing for computing Fibonacci spirals as shown on p. 172).

CHAPTER 16

1. The various physical cave structures, such as the gypsum chandeliers and aragonite bushes, have been found in caves in the U.S. but do not necessarily reside in La Pileta, the Number Cave.

CHAPTER 19

1. One of the most enigmatic examples of the intertwining of mathematics and mysticism occurs in the final chapters of Thomas Pynchon's novel *Gravity's Rainbow* where I first learned about Kabala. Pynchon explains that the rocket countdown sequence which precedes a launching—"10, 9, 8, 7, 6, 5, 4, 3, 2, 1, 0"—was invented by Fritz Lang in the 1929 Ufa film, *Die Frau in Mond.* Lang added it to the launch scene to heighten the suspense.

   A Kabalist in Pynchon's novel explains that during the creation of the universe, God sent a pulse of energy into the void. The pulse branched into 10 distinct spheres or aspects, corresponding to the numbers 1 through 10. These 10 aspects are known as the sephiroth. To return to God, the soul must negotiate each of the sephiroth, from 10 back to 1. Pynchon tells us that Kabalists armed with magic and faith attempt to conquer the sephiroth. In fact, most of the Kabalists' secrets have to do with successfully making the trip.

   A few pages from the end of *Gravity's Rainbow*, Pynchon's main character explains that the sephiroth fall into a pattern called the Tree of Life, the body of God. There are 22 paths drawn among the 10 spheres. Each path corresponds to a letter of the Hebrew alphabet and also to one of the major arcana cards of the Tarot. Pynchon enigmatically concludes *Gravity's Rainbow* with the idea that the rocket countdown sequence conceals the Tree of Life. Pynchon says, "Although the Rocket countdown appears to be serial, it actually conceals the Tree of Life, which must be perceived all at once, together, in parallel."

CHAPTER 21

1. David Glass from Lexington, Kentucky, writes to me: "It is interesting that the Russian empire experienced such a variety of difficulties in a period of less than two decades. I'm sure seers of the time must have thought the Tunguska blast a foretelling of the coming revolution and the toppling of the Tsar. The Tunguska blast could have been considered very prophetic if Alexis Romanov had been born a few years later. As it was, Alexis Romanov had already been diagnosed as a hemophiliac by the time of the blast, and this was one of the causes of the fall of the empire."

2. Those of you who are mathematicians may be interested in how I created the distribution of crater sizes shown on p. 217. To approximate a $1/x^2$ distribution, I used a "one-sided" Cauchy distribution: $f(x) = 2/\pi\beta(1+(x/\beta)^2)$. The desired one-sided Cauchy distribution can be created from uniform random variates U on (0,1) (as are produced by typical random number generators) via the Czarnecki transformation: $X = \beta \tan(U\pi/2)$. In this formula, β determines the location of the distribution's "knee." For $x > \beta$, the one-sided Cauchy probability density function is proportional to $1/x^2$. For $0 < x < \beta$ one-sided Cauchy is approximately flat. See the Appendix program code for more details.

3. A *meteor* is the light streak or "shooting star" which can be seen in the sky. The cause of the streak is a *meteorite*, a solid object which plunges to earth. A *meteoroid* is the object while it is still out in space. An *asteroid* is a large meteoroid. Sometimes asteroids are called *planetoids*. *Meteorism* is a medical term for rapid belly ballooning caused by a dead bowel.

4. Orthodox Moslems believe that the Kaaba was rebuilt 10 times. The first Kaaba was erected at the dawn of history by angels from heaven. The second was built by Adam. The third by his son Seth. The fourth by Abraham, the seventh by Qusay, the eighth by the Quraish leaders in Mohammed's

lifetime (605), the ninth and tenth by Moslem leaders in 681 and 696. The tenth building is substantially the Kaaba of today.

In A.D. 629, Mohammed reverently touched the Black Stone with his staff. A year later, he destroyed the idols in and around the Kaaba, but spared the Black Stone, and sanctioned the kissing of it. He proclaimed Mecca the Holy City of Islam, and decreed that no unbeliever should be allowed to stand on its holy soil.

Since the days of Mohammed, the Black Stone has suffered much. For example, in 680, the Moslems of Kufa told Husein in Mecca they would make him caliph if he would make their city his capital. When Husein's followers chose to fight, Husein and his brothers and sons were slaughtered while the women and children looked on in horror. When Husein's severed head was brought to Obeidallah, governor of Kufa, he poked at it with his staff. "Carefully," one of his officers protested, "he was the grandson of the Prophet. By Allah! I have seen his lips kissed by the blessed mouth of Mohammed."

The revolt against Husein's followers continued. A few years after the death of Husein, rocks from Syrian catapults fell upon the Kaaba and split the Black Stone into three pieces. (The Syrian troops were led by Yezid, the caliph of the Moslem empire.) The Kaaba caught fire, and was burned to the ground in 683.

The pre-Islamic desert Arabs had interesting customs. For example, these hardy folk tied camels, without food, to their graves so that the camels would follow them to the next world. In this way, the Arabs wouldn't have to walk on foot in paradise. Historian Will Durant notes that in pre-Moslem days, there were several idols within the Kaaba that represented gods. One idol was called Allah, the tribal god of the Quraish. Three others were Allah's daughters.

5. Doomsday (or near-Doomsday) scenarios by comets or asteroids are common in the science fiction literature. A personal favorite is Arthur C. Clarke's *Hammer of God* (1993). Other interesting examples are Gregory Benford's and William Rotsler's *Shiva Descending* (1980), Larry Niven's and Jerry Pournelle's *Lucifer's Hammer*, James Blish's and Norman Knight's *A Torrent of Faces* (1967), and Charles Diffin's short story "The Hammer of Thor" (1932).

POSTSCRIPT 1

1. To get a feel for Usenet's popularity, consider that over 40,000 messages are sent each day to thousands of newsgroups in Usenet. The most popular newsgroup, news.announce.newusers, has more than 800,000 readers. Note that a newsgroup called "alt.fan.cliff-pickover" exists for discussions on topics and problems discussed in my books.

Some define "Usenet" as the set of people (not computers) who exchange notes and news articles tagged with one or more universally recognized labels signifying a particular "newsgroup." There are many hundreds of news groups on topics ranging from bicycles, to physics, to music. Usenet started out at Duke University around 1980 as a small network of UNIX machines. Today there is no UNIX limitation, and there are versions of the news-exchange programs that run on computers ranging from DOS PCs to mainframes. Most Usenet sites are at universities, research labs, or other academic and commercial institutions. The largest concentrations of Usenet sites outside the U.S. are in Canada, Europe, Australia, and Japan. Interestingly, the Internet (a major computer network linking millions of machines around the world) is growing at a phenomenal rate. Between January 1993 and January 1994, the number of connected machines grew from 1,313,000 to 2,217,000. Over 70 countries have full Internet connectivity, and about 150 have at least some electronic mail services. In 1994, there were about 20–25 million users of the Internet. (Source: Goodman et al. (1994). The global diffusion of the Internet. *Commun. ACM* 37(8):27.)

References

GENERAL READING

Cohen, D. (1983) *Waiting for the Apocalypse*. Prometheus Books: New York.

Davies, P. (1992) *The Mind of God*. Touchstone/Simon and Schuster: New York.

Davis, P. and Hersh, R. (1981) *The Mathematical Experience*. Houghton Mifflin Company: Boston.

Davis, P. and Hersh, R. (1986) *Descartes' Dream*. Harcourt Brace Jovanovich: New York.

Ellis, K. (1978) *Number Power*. St. Martin's Press: New York.

Gardner, M. (1992) *A Walk on the Wild Side*. Prometheus Books: New York.

Gorman, P. (1979) *Pythagoras: A Life*. Routledge and Kegan Paul: Boston.

Guthrie, K. (1991) *The Pythagorean Sourcebook and Library*. Phanes Press: Grand Rapids, Michigan.

Kappraff, J. (1991) *Connections*. McGraw-Hill: New York.

McLeish, J. (1991) *The Story of Numbers*. Fawcette: New York.

Russell, B. (1945) *A History of Western Philosophy*. Simon and Schuster: New York.

Schimmel, A. (1993) *The Mystery of Numbers*. Oxford University Press: New York.

CHAPTER 1

Belier, A. (1966) *Recreations in the Theory of Numbers*. Dover: New York.

Davis, P. and Hersh, R. (1981) *The Mathematical Experience*. Houghton Mifflin Company: Boston.

Ellis, K. (1978) *Number Power*. St. Martin's Press: New York.

Guy, R. (1994) Every number is expressible as the sum of how many polygonal numbers? *American Mathematics Monthly*. 101(2): 169–172.

Heath, T. (1963) *Greek Mathematics*. Dover: New York.

Kordemsky, B. (1972) *The Moscow Puzzles*. Dover: New York.

McLeish, J. (1991) *The Story of Numbers*. Fawcette: New York.

Nelsen, R. (1991) Proof without words: Sum of reciprocals of triangular numbers. *Mathematics Magazine* 64(3): 167.

Nelsen, R. (1993) *Proofs without Words: Exercises in Visual Thinking*. Mathematics Association of America: New York.

Nelsen, R. (1994) Proof without words: A triangular identity. *Mathematics Magazine*. 67(4): 293.

Sarton, G. (1952) *A History of Science*. Harvard University Press: Massachusetts.

Schimmel, A. (1993) *The Mystery of Numbers*. Oxford University Press: New York.

Schroeder, M. (1984) *Number Theory in Science and Communication*. Springer: New York.

Wells, D. (1987) *The Penguin Dictionary of Curious and Interesting Numbers*. Penguin: New York. (Many of the interesting triangular number formulas come from this book.)

CHAPTER 3

Deakin, M. (1994) Hypatia and her mathematics. *American Mathematical Monthly*. 101(3): 234–243.

Gorman, P. (1979) *Pythagoras: A Life*. Routledge and Kegan Paul: Boston.

Miller, W. (1993) Proof without words: Sum of pentagonal numbers. *Mathematics Magazine*. 66(5): 325.

Pickover, C. (1992) Chapter 47, Undulating undecamorphic and undulating pseudofareymorphic integers. In *Computers and the Imagination*. St. Martin's Press: New York. Also: Chapter 37, On the existence of cakemorphic integers (describes square pyramidal numbers). Also: Chapter 36, Infinite sequences in centered hexamorphic numbers (describes hexagonal numbers and centered hexagonal numbers).

Schimmel, A. (1993) *The Mystery of Numbers*. Oxford University Press: New York.

CHAPTER 4

Kroon, R., Sancton, T. and Scott, G. (1994) In the reign of fire. *Time*. October 17. 59–60. (Describes the Order of the Solar Temple.)

Pearl, R. and Reed, L. (1920) On the rate of growth of the population of the United States since 1790 and its mathematical representation. *Proceedings of the National Academy of Sciences*. 6(6): 275–288.

von Foerster, H., Mora, P., and Amiot, L. (1960) Doomsday: Friday 13 November A.D. 2026. *Science*. 132: 1291–1295.

The following are controversial responses to the von Foerster paper:

Robertson, J., Bond, V., and Cronkite, E. (1961) Doomsday (Letter to the Editor) *Science*. 133: 936–937.

Hutton, W. (1961) Doomsday (Letter to the Editor) *Science*. 133: 937–939.

Howland, W. (1961) Doomsday (Letter to the Editor) *Science*. 133: 939–940.

Shinbrot, M. (1961) Doomsday (Letter to the Editor) *Science*. 133: 940–941.

von Foerster, H., Mora, P., and Amiot, L. (1961) Doomsday (Letter to the Editor) *Science*. 133: 941–952.

Serrin, J. (1975) Is "Doomsday" on Target? *Science*. 189: 86–88.

CHAPTER 5

Beiler, A. (1964) *Recreations in the Theory of Numbers*. Dover: New York. (Contains much information on Pythagorean triangles.)

Dudley, U. (1992) *Mathematical Cranks*. Mathematical Association of America: Washington, D.C.

Schimmel, A. (1993) *The Mystery of Numbers*. Oxford University Press: New York.

CHAPTER 6

Ellis, K. (1978) *Number Power*. St. Martin's Press: New York.
Wells, D. (1987) *The Penguin Dictionary of Curious and Interesting Numbers*. Penguin: New York.

CHAPTER 7

Beiler, A. (1964) *Recreations in the Theory of Numbers*. Dover: New York.
Spencer, D. (1982) *Computers in Number Theory*. Computer Science Press: New York.

CHAPTER 8

Mott-Smith, G. (1954) Turks and Christians. In *Mathematical Puzzles*. Dover: New York. (p. 94)

CHAPTER 9

Gardner, M. (1981) *Science: Good, Bad and Bogus*. Prometheus Books: Buffalo, New York. (Chapter 3 describes the work of Ramon Lull in fascinating detail and is the primary source for my information on Lull in this chapter. Gardner's book is highly recommended.)
Reichardt, J. (1969) *Cybernetic Serendipity: The Computer and the Arts*. Praeger: New York. (This book has a chapter on Japanese haiku, computer texts, high-entropy essays, fairytales, and fake physics essays.)
Kress, N. (1994) An Untitled Column. *Writer's Digest*. December, 8–10. (Describes the Lullian generation of book titles.)
Kurzweil, R. (1990) *The Age of Intelligent Machines*. MIT Press: Cambridge, Massachusetts. (This book contains information on pattern recognition, the science of art, computer-generated poetry, and artificial intelligence.)
Pickover, C. (1992) *Computers and the Imagination*. St. Martin's Press: New York. (Describes computer poetry.)
Pickover, C. (1994) *Mazes for the Mind: Computers and the Unexpected*. St. Martin's Press: New York. (Describes the Lullian approach for creating inventions.)
Pickover, C. (1995) *Chaos in Wonderland: Visual Adventures in a Fractal World*. St. Martin's Press: New York.

CHAPTER 10

Alvarez, L.W., Alvarez, W., Asaro, F., Michel, H.V. (1980) Extraterrestrial cause for the Cretaceous-Tertiary extinction. *Science*. 208, 1095–1108.
Alvarez, W., Asaro, F., Michel, H.V., Alvarez, L.W. (1982) Iridium anomaly approximately synchronous with terminal Eocene extinctions. *Science*. 216, 886–888.
Beatty, J. (1994) Secret impacts revealed. *Sky and Telescope*. February. 26–27. (Describes atmospheric impacts recently revealed by the U.S. Department of Defense.)
Cohen, D. (1983) *Waiting for the Apocalypse*. Prometheus Books: New York.
Davies, P. (1994) *The Last Three Minutes*. Basic Books: New York. (Discusses the Swift-Tuttle comet and Nemesis or the Death Star.)
Davis, M., Hut, P. and Muller, R. (1984) Extinction of species by periodic comet showers. *Nature*. 308: 715–717.

Hoffman, A. (1985) Patterns of family extinction depend on definition and geological timescale. *Nature*. 315: 659–662.

Moore, P., Hunt, G., Nicolson, I., and Cattermole, P. (1990) *The Atlas of the Solar System*. Crescent Books/Crown Publishers: New York.

Rampino, M. and Stothers, R. (1984) Terrestrial mass extinctions, cometary impacts and the Sun's motion perpendicular to the galactic plane. *Nature*. 308: 709–712.

Raup, D.M. and Sepkoski, J. (1984) Periodicity of extinctions in the geological past. *Proceedings of the National Academy of Science*. 81: 801–805.

Schwartz, R.D. and James, P.B. (1984) Periodic mass extinctions and the Sun's oscillation about the galactic plane. *Nature*. 308, 712–713.

Whitmire, D. and Jackson, A. (1984) Are periodic mass extinctions driven by a distant solar companion? *Nature*. 308: 713–715.

CHAPTER 11

Daniel, G. (1980) Megalithic monuments. *Scientific American*. 243(1): 78–90.

Hawkins, G. (1964) Stonehenge—a neolithic computer. *Nature* 202: 1258–1261.

Hawkins, G. (1973) *Beyond Stonehenge*. Harper & Row: New York.

Hoyle, F. (1966) Stonehenge—an eclipse predictor. *Nature*. 211(5048): 454–458.

CHAPTER 12

Gardner, M. (1992) The great *Urantia Book* mystery. In *A Walk on the Wild Side*, Prometheus Books: New York. (Most of my information on Urantia comes from this wonderful book.)

Gardner, M. (1995) *Urantia: The Great Cult Mystery*. Prometheus Books: New York.

CHAPTER 13

Penrose, R. (1989) *The Emperor's New Mind: Concerning Computers, Minds, and the Laws of Physics*. Oxford University Press: Oxford.

Yates, F. (1964) *Giordano Bruno and the Hermetic Tradition*. University of Chicago: Chicago.

The following is a list of fractal resources sorted into various categories.

Books

1. Anthony, P. (1992) *Fractal Mode*. Ace: New York. (Science fiction.)
2. Barnsley, M. (1988) *Fractals Everywhere*. Academic Press: New York.
3. Batty, M., and Longley, P. (1994) *Fractals Cities: A Geometry of Form and Function*. Academic Press: California. (Topic: the development and use of fractal geometry for understanding and planning the physical form of cities; shows how fractals enable cities to be simulated through computer graphics.)
4. Briggs, J. (1992) *Fractals*. Simon and Schuster: New York.
5. Falconer, K. (1990) *Fractal Geometry*. Wiley: New York.
6. Feder, J. (1988) *Fractals*. Plenum: New York.
7. Fischer, P., Smith, W. (1985) *Chaos, Fractals, and Dynamics*. Marcel Dekker, Inc.: New York.
8. Glass, L., Mackey, M. (1988) *From Clocks to Chaos: The Rhythms of Life*. Princeton University Press: New Jersey.

9. Gleick, J. (1987) *Chaos*. Viking: New York.

10. Kaye, B. (1989) *A Random Walk Through Fractal Dimensions*. VCH Publishers: New York.

11. Lauwerier, H. (1990) *Fractals*. Princeton University Press, Princeton, New Jersey.

12. Mandelbrot, B. (1984) *The Fractal Geometry of Nature*. Freeman: New York.

13. Moon, F. (1987) *Chaotic Vibrations*. Wiley: New York.

14. Peak, D. and Frame, M. (1994) *Chaos Under Control: The Art and Science of Complexity*. Freeman: New York.

15. Peitgen, H., Richter, P. (1986) *The Beauty of Fractals*. Springer: Berlin.

16. Peitgen, H., Saupe, D. (1988) *The Science of Fractal Images*. Springer: Berlin.

17. Pickover, C. (1990) *Computers, Pattern, Chaos and Beauty*. St. Martin's Press: New York.

18. Pickover, C. (1991) *Computers and the Imagination*. St. Martin's Press: New York.

19. Pickover, C. (1992) *Mazes for the Mind: Computers and the Unexpected*. St. Martin's Press: New York.

20. Pickover, C. (1994) *Chaos in Wonderland: Visual Adventures in a Fractal World*. St. Martin's Press: New York.

21. Pickover, C. (1995) *Keys to Infinity*. Wiley: New York.

22. Pickover, C. (1995) *The Pattern Book: Fractals, Art and Nature*. World Scientific: New Jersey.

23. Pickover, C. (1996) *Fractal Horizons: The Future Use of Fractals*. St. Martin's Press: New York.

24. Peterson, I. (1990) *Islands of Truth*. Freeman: New York.

25. Rietman, E. (1989) *Exploring the Geometry of Nature*. Windcrest: Pennsylvania.

26. Sprott, C. (1993) *Strange Attractors: Creating Patterns in Chaos*. M&T Books: New York. (A book by the guru of strange attractors and their graphics.)

27. Schroeder, M. (1984) *Number Theory in Science and Communication*. Springer: New York. (This book is recommended highly. An interesting book by a fascinating author.)

28. Schroeder, M. (1991) *Fractals, Chaos, Power Laws*. Freeman: New York.

29. Stewart, I. (1980) *Does God Play Dice?* Basil Blackwell: New York.

30. Stevens, C. (1989) *Fractal Programming in C*. M&T Books: New York.

31. Wegner, T., Peterson. M. (1991) *Fractal Creations*. Waite Group Press: California.

32. Wegner, T., Tyler, B., Peterson, M. and Branderhorst, P. (1992) *Fractals for Windows*. Waite Group Press: California.

33. West, B. (1990) *Fractal Physiology and Chaos in Medicine*. World Scientific: New Jersey.

Newsletters, Organizations, and Related

1. *AMYGDALA*, a newsletter on fractals. Write to AMYGDALA, Box 219, San Cristobal, New Mexico 87564 for more information.

2. *ART MATRIX*, creator of beautiful postcards and videotapes of exciting mathematical shapes. Write to ART MATRIX, PO Box 880, Ithaca, New York 14851 for more information.

3. *Bourbaki Software*: *FracTools* (fractal generation software), *A Touch of Chaos* (fractal screensaver for windows), *FracShow CD* (fractal slide shows on CD-ROM for DOS and Windows). Contact: Bourbaki Inc, PO Box 2867, Boise, ID 83701.

4. *Chaos Demonstrations*, a program which contains examples of Julia sets and over twenty other types of chaotic systems. This peer-reviewed program won the first annual *Computers in Physics* contest for innovative educational physics software. It is published by Physics Academic Software and is available from The Academic Software Library, Box 8202, North Carolina State University, Raleigh, NC 27695-8202. Chaos Demonstrations is by J. C. Sprott and G. Rowlands, and it requires an IBM PC or compatible computer.

5. "Chaos and Graphics Section" of *Computers and Graphics*. (This section is devoted to fractals and chaos in art and science.) Publishing, subscription, and advertising office for *Computers and Graphics*: Elsevier Science, Inc., 660 White Plains Road, Tarrytown, New York 10591-5153. Email: ESUK.USA@ELSEVIER.COM

6. *Fractal Calendar.* Address inquiries to J. Loyless, 5185 Ashford Court, Lilburn, Georgia 30247.

7. *Fractal Discovery Laboratory.* Designed for a science museum or school setting. Entertaining for a 4-year-old, and fascinating for the mathematician. Earl Glynn, Glynn Function Study Center, 10808 West 105th Street, Overland Park, KS 66214-3057.

8. Fractal Postcards. The Mathematical Association, 259 L Leicester, LE2 3BE U.K. *Fractal Report*, a newsletter on fractals. Published by J. de Rivaz, Reeves Telecommunications Lab. West Towan House, Porthtowan, Cornwall TR4 8AX, U.K.

9. *HOP—Fractals in Motion.* Software which produces a large variety of novel images and animations. HOP features Fractint-like parameter files, GIF read/write, MAP palette editor, a screensaver for DOS, Windows, and OS/2, and more. Math coprocessor (386 and above) and SuperVGA required. The program is written by Michael Peters and Randy Scott. $30 shareware. Download locations: Compuserve GRAPHDEV Forum, Lib 4 (HOPZIP.EXE), The WELL ibmpc/graphics (HOPZIP.EXE), rever.nmsu.edu under /pub/hop, slopoke.mlb.semi.harris.com, ftp.uni-heidelberg.de under /pub/msdos/graphics, spanky.triumf.ca [128.189.128.27] under [pub.fractals.programs.ibmpc], HOP WWW page: http://rever.nmsu.edu/~ras/hop. Subscriptions and information requests for the HOP mailing list should be sent to: hop-request@acca. nmsu.edu. To subscribe to the HOP mailing list, simply send a message with the word "sub-scribe" in the subject: field. For information, send a message with the word "INFO" in the subject: field.

10. The Great Media Company, PO Box 598, Nicasio, California 94946. This fine company distributes books, videos, prints, and calendars.

11. *Recreational and Educational Computing Newsletter.* Dr. Michael Ecker, 909 Violet Terrace, Clarks Summit, PA 18411.

12. *Strange Attractions.* A store devoted to chaos and fractals (fractal art work, cards, shirts, puzzles, and books). For more information, contact: *Strange Attractions*, 204 Kensington Park Road, London W11 1NR England.

13. *StarMakers Rising* (Fractal posters), 6801 Lakeworth Road, Suite 201, Lakeworth, FL 33467.

14. *YLEM—Artists using science and technology.* This newsletter is published by an organization of artists who work with video, ionized gases, computers, lasers, holograms, robotics, and other nontraditional media. It also includes artists who use traditional media but who are inspired by images of electromagnetic phenomena, biological self-replication, and fractals. Contact: YLEM, Box 749, Orinda, CA 94563.

Recent Unconventional Articles

(This list is not meant to be comprehensive. It is included for researchers in the field who may not be aware of some of the following unusual applications. Some of the more "offbeat" topics are listed first.)

1. Taylor, C. (1990) Condoms and cosmology: The "fractal" person and sexual risk in Rwanda. *Social Science and Medicine.* 31(9): 1023–1028.

2. Entsminger, G. (1989) Stochastic fiction—fiction from fractals. *Micro Cornucopia.* 49: 96.

3. Pickover, C. (1993) Fractal fantasies. *BYTE.* March. p. 256. (Describes a game played on a fractal playing board.)

4. Lakhtakia, A. (1990) Fractals and The Cat in the Hat. *Journal of Recreational Math.* 23(3): 161–164.

5. Nottale, L. (1991) The fractal structure of the quantum space-time. In Heck, A. and Perdang, J. *Applying Fractals in Astronomy.* Springer: New York.

6. Landini, G. (1991) A fractal model for periodontal breakdown in periodontal disease. *J. Periodontal Res.* 26: 176–179.

7. Cutting, J., and Garvin, J. (1987) Fractal curves and complexity. *Perception and Psychophysics.* 42: 365–370.

8. Batty, M. and Longley, P. (1987) Urban shapes as fractals. *Area*. 19: 215–221.

9. Fogg, L. (1989) PostScriptals: Ultimate fractals via postscript. *Micro Cornucopia*. 49: 16–22. (Discusses the "ultimate" fractal at a resolution of 2540 dots per inch. Also challenges the reader to beat this "world's highest resolution fractal.")

10. Dixon, R. (1992) The pentasnow gasket and its fractal dimension. In *Fivefold Symmetry*. Hargittai, I., ed. World Scientific: New Jersey.

11. Clarke, A. (1989) *Ghost from the Grand Banks*. Bantam: New York. (A female character becomes insane after exploring the Mandelbrot set.)

12. Pickover, C. (1995) Is the fractal golden curlicue cold? *The Visual Computer*. 11(6): 309–312.

13. Lakhtakia, A., Messier, R., Vasandara, V., Varadan, V. (1988) Incommensurate numbers, continued fractions, and fractal immittances. *Zeitschrift für Naturforschung* 43A: 943–955.

14. Pool, R. (1990) Fractal Fracas. *Science*. 249: 363–364. ("The math community is in a flap over the question of whether fractals are just pretty pictures or more substantial tools.")

15. Batty, M., and Longley, P., Fotheringham, A. (1989) Urban growth and form: scaling, fractal geometry and diffusion-limited aggregation. *Environment and Planning* 21:1447–1472.

16. Hsu, K., Hsu A. (1990) Fractal geometry of music. *Proceedings of the National Academy of Science*. 87(3):938–941.

Other Fractal Articles

1. Aqvist, J., Tapia, O. (1987) Surface fractality as a guide for studying protein–protein interactions. *Journal of Molecular Graphics*. 5(1): 30–34.

2. Barnsley, M., Sloan, A. (1987) Chaotic compression (a new twist on fractal theory speeds complex image transmission to video rates). *Computer Graphics World*. November. 107–108.

3. Batty, M. (1985) Fractals—geometry between dimensions. *New Scientist*. April. 31–40.

4. Boyd, D. (1973) The residual set dimensions of the Apollonian packing. *Mathematika* 20: 170–74. (Very technical reading)

5. Brooks, R., Matelski, J. P. (1981) The dynamics of 2-generator subgroups of PSL(2,C). In *Riemann Surfaces and Related Topics: Proceedings of the 1978 Stony Brook Conference*. Kyra, I. and Maskit, B. (eds.) Princeton University Press: New Jersey. (Note: This 1978 paper contains computer graphics and mathematical descriptions of both Julia and Mandelbrot sets.)

6. Casey, S. (1987) Formulating fractals. *Computer Language*. 4(4): 28–38.

7. Collins, J.J. and De Luca, C.J. (1993) Open-loop and closed-loop control of posture: A random-walk analysis of center-of-pressure trajectories. *Experimental Brain Research*. 95: 308–318.

8. Collins, J.J. and De Luca, C.J. (1994) Random walking during quiet standing. *Physical Review Letters*. 73: 764–767. (The Collins papers deal with fractal-like measures and the postural control system.)

9. Devaney, R., Krych, M. (1984) Dynamics of exp(z). *Ergodic Theory & Dynamical Systems*. 4: 35–52.

10. Dewdney, A. K. (1985) Computer Recreations. *Scientic American*. 253: 16–24.

11. Douady, A., Hubbard, J. (1982) Iteration des polynomes quadratiques complexes. *Comptes Rendus* (Paris) 2941: 123–126.

12. Family, F. (1988) Introduction to droplet growth processes: Simulation theory and experiments. In *Random Fluctuations and Pattern Growth: Experiments and Models*. Stanley, H., Ostrowsky, N. (eds.) Kluwer: Boston.

13. Fatou, P. (1919/1920) Sur les equations fonctionelles. *Bull. Soc. Math. Fr.* 47: 161–271.

14. Fournier, A., Fussel, D., Carpenter, L. (1982) Computer rendering of stochastic models. *Communications of the ACM* 25: 371–378.(How to create natural irregular objects.)

15. Gardner, M. (1978) White and brown music, fractal curves, and 1/f fluctuations. *Scientific American*. April 16–31.

16. Gardner, M. (1976) In which monster curves force redefinition of the word curve. *Scientific American*. December 235: 124–133.

17. Gordon, J., Goldman, A., Maps, J. (1986) Superconducting-normal phase boundary of a fractal network in a magnetic field. *Phys. Rev. Let.* 56: 2280–2283.

18. Grebogi, C., Ott, E., Yorke, J. (1985) Chaos, strange attractors, and fractal basin boundaries in nonlinear dynamics. *Science* 238: 632–637. (A great overview with definitions of terms used in the chaos literature.)

19. Grebogi, C., Ott, E., Yorke, J. (1985) Attractors on an N-Torus: Quasiperiodicity versus chaos. *Physica* 15D: 354–373. (Contains some gorgeous diagrams of dynamical systems).

20. Hirsch, M. (1989) Chaos, rigor, and hype. *Mathematical Intelligencer.* 11(3): 6–9. (Pages 8 and 9 include James Gleick's response to the article.)

21. Holter, N., Lakhtakia, A., Varadan, V., Vasundara, V., Messier, R. (1986) On a new class of planar fractals: the Pascal-Sierpinski gaskets. J. Phys. A: Math. Gen. 19: 1753–1759.

22. Julia, G. (1918) Memoire sur l'iteration des fonctions rationnelles. *J. Math. Pure Appl.* 4: 47–245.

23. Kadanoff, L. (1986) Fractals: Where's the physics? *Physics Today.* February: 6–7.

24. La Brecque, M. (1985) Fractal Symmetry. *Mosaic.* 16: 10–23.

25. Lakhtakia, A., Vasundara, V., Messier, R., Varadan, V. (1987) On the symmetries of the Julia sets for the process $z \to z^p + c$. *J. Phys. A: Math. Gen.* 20: 3533–3535.

26. Lakhtakia, A., Vasundara, V., Messier, R., Varadan, V. (1987) The generalized Koch Curve. *J. Phys. A: Math. Gen.* 20: 3537–3541.

27. Lakhtakia, A., Vasundara, V., Messier, R., Varadan, V. (1986) Self-similarity versus self-affinity: the Sierpiñski gasket revisited. *J. Phys. A: Math. Gen.* 19: L985–L989.

28. Lakhtakia, A., Vasundara, V., Messier, R., Varadan, V. (1988) Fractal sequences derived from the self-similar extensions of the Sierpinski gasket. *J. Phys. A: Math. Gen.* 21: 1925–1928.

29. Mandelbrot, B. (1983) On the quadratic mapping $z \to z^2 - \mu$ for complex μ and z: The fractal structure of its M set, and scaling. *Physica.* 17D: 224–239.

30. Musgrave, K. (1989) The synthesis and rendering of eroded fractal terrains. *Computer Graphics (ACM-SIGGRAPH).* July 23(3): 41–50.

31. Norton, A. (1982) Generation and display of geometric fractals in 3-D. *Computer Graphics (ACM-SIGGRAPH).* 16: 61–67.

32. Peterson, I. (1984) Ants in the labyrinth and other fractal excursions. *Science News.* 125: 42–43.

33. Pickover, C. (1988) Symmetry, beauty and chaos in Chebyshev's Paradise. *The Visual Computer: An International Journal of Computer Graphics.* 4:142–147.

34. Pickover, C. (1987) Biomorphs: computer displays of biological forms generated from mathematical feed back loops, *Computer Graphics Forum.* 5(4): 313–316.

35. Pickover, C. (1995) Synthesizing extraterrestrial terrain, *IEEE Computer Graphics and Applications*, 15: 18–21.

36. Phipps, T. (1985) Enhanced fractals. *Byte.* March 21–23.

37. Robinson, A. (1985) Fractal fingers in viscous fluids. *Science.* 228: 1077–1080.

38. Schroeder, P. (1986) Plotting the Mandelbrot Set. *Byte.* December 207–211.

39. Schroeder, M. (1982) A simple function. *Mathematical Intelligencer.* 4: 158–161.

40. Sorenson, P. (1984) Fractals. *Byte.* Sept. 9: 157–172 (A fascinating introduction to the subject.)

41. Symmetries and Asymmetries. (1985) *Mosaic.* Volume 16, Number 1, January/February. (An entire issue on the subject of fractals, symmetry, and chaos. *Mosaic* is published six times a year as a source of information for scientific and educational communities served by the National Science Foundation, Washington, D.C. 20550).

42. Thomsen, D. (1982) A place in the sun for fractals. *Science News.* 121: 28–32.

43. Thomsen, D. (1980) Making music–fractally. *Science News* March 117:189–190.

44. Ushiki, S. (1988) Phoenix. *IEEE Transactions on Circuits and Systems.* July 35(7): 788–789.

45. Voss, R. (1985) Random fractal forgeries. In *Fundamental Algorithms in Computer Graphics.* R. Earnshaw, ed. Springer-Verlag: Berlin, pp. 805–835.

46. West, S. (1984) The new realism. *Science.* 84, 5: 31–39.

47. West, B., Goldberger, A. (1987) Physiology in fractal dimensions. *American Scientist*. 75: 354–365. (This article describes the fractal characterization of the lungs' bronchial tree, the Weierstrass function, the fractal geometry of the heart, and "fractal time.")

CHAPTER 14

Ascher, M., and Ascher, R. (1981) *The Code of the Quipu: A Study in Media, Mathematics and Culture*. Ann Arbor, MI: University of Michigan Press.

Ascher, M. (1991) *Ethnomathematics: A Multicultural View of Mathematical Ideas*. Pacific Grove, CA: Brooks/Cole Publishing Co.

Ascher, M. (1992) Before the conquest. *Mathematics Magazine*. Oct 65(4): 211–218.

Conklin, W. (1982) The information system of middle horizon quipus. In *Ethnoastronomy and Archaeoastronomy in the American Tropics*. Aveni, A. and Urton, G., eds. The New York Academy of Sciences: New York, pp. 261–281.

Hawkins, G. (1973) *Beyond Stonehenge*. Harper and Row: New York.

McLeish, J. (1991) *The Story of Numbers*. Fawcette: New York.

Zaslavsky, C. (1973) *Africa Counts: Number and Pattern and African Culture*. Brooklyn, NY: Lawrence Hill Press.

CHAPTER 15

Barret, A., Mackay, A. (1987) *Spatial Structure and the Microcomputer*. Macmillan: New York.

Boles, M., Newman, R. (1990) *Universal Patterns*. Pythagorean Press: Bradford, Massachusetts.

Gardner, M. (1994) The cult of the Golden Ratio. *Skeptical Inquirer*. Spring 18(3): 243–247.

MacEoin, D. (1992) *The Sources for Early Babi Doctrine and History*. E. J. Brill: New York.

Markowsky, G. (1992) Misconceptions about the Golden Ratio. *College Mathematics Journal*. January. 23: 2–19.

Schroeder, M. (1986) *Number Theory in Science and Communication*. Springer: Berlin. (A goldmine of valuable information.)

Stewart, I. (1995) Daisy, daisy, give me your answer do. *Scientific American*. January. 272(1): 96–99.

Wells, D. (1987) *The Penguin Dictionary of Curious and Interesting Numbers*. Penguin: New York.

CHAPTER 16

Hawkins, G. (1973) *Beyond Stonehenge*. Harper and Row: New York.

CHAPTER 17

Pickover, C. (1992) *Mazes for the Mind: Computers and the Unexpected*. St. Martin's Press: New York.

Pickover, C. (1993) Lava lamps in the 21st century. *Visual Computer*. Dec 10(3): 173–177.

Vichniac, G. (1986) Cellular automata models of disorder and organization. In *Disordered Systems and Biological Organization*. Bienenstock., E., Soulie, F., and Weisbuch, G., eds. Springer: New York.

CHAPTER 18

Hawkins, G. (1973) *Beyond Stonehenge*. Harper and Row: New York.

CHAPTER 19

Ellis, K. (1978) *Number Power*. St. Martin's Press: New York.

Flam, F. (1994) Theorists make a bid to eliminate black holes. *Science*. 266(5193): 1945.

Horgan, J. (1995) Bashing black holes: theorists twist relativity to eradicate an astronomical anomaly. *Scientific American*. July 273(1): 16.

Lemonick, M. and Nash, M. (1995) Unraveling universe. *Time*, March 6, 145(9): 77–84.

Pynchon, T. (1973) *Gravity's Rainbow*. Bantam: New York.

CHAPTER 21

Cherrington, E. (1969) *Exploring the Moon through Binoculars and Small Telescopes*. Dover: New York.

Cohen, D. (1983) *Waiting for the Apocalypse*. Prometheus Books: New York.

Dyson, F. (1979) Time without end: physics and biology in an open universe. *Reviews of Modern Physics*. 51(3).

Gallant, R. (1994) Journey to Tunguska. *Sky & Telescope*. June. 87: 38–43.

Marsden, B. (1993) Comet Swift-Tuttle: Does it threaten Earth? *Sky & Telescope*. January. 85: 16–19.

Morrison, D. and Chapman, C. (1990) Target Earth: It *will* happen. *Sky & Telescope*. March. 79: 261–262.

Weissman, R. (1990) Are periodic bombardments real? *Sky & Telescope*. March. 79: 266–270.

POSTSCRIPT 1

Dauben, J. (1979) *Georg Cantor*. Harvard University Press: Cambridge (Chapter 6. pp. 246, 295).

Gunn, J. (1988) *The New Encyclopedia of Science Fiction*. Viking: New York.

Hick, J. (1978) *Evil and the God of Love*. Harper and Row: New York.

Pickover, C. and Webb, D. (1995) To the Valley of the Seahorses. In *Keys to Infinity* (Pickover). Wiley: New York (Chapter 6, pp. 47–58).

Tipler, F. (1994) *The Physics of Immortality*. Doubleday: New York.

Wang, Hao (1987) Reflections on Kurt Goedel. MIT Press: Mass. (p. 195)

Index

Abacus, 179
Abdu'1-Baha, 40, 70
Abundant numbers, 87
Allah, 76, 81, 275
Alvarez, Luis, 128
Amanesis, 49
Amicable chain, 90
Amicable numbers, 87, 89, 91, 260
Anselm, 200, 232
Apocalypse, 24–25, 42, 52, 57, 65, 131,
    133, 207–224, 247
Apollo, 51
Apollo, Temple of, 29
Aquinas, Thomas, 198–199
Arabian Nights, 98
Aristotle, 24
Aristoxenus, 49
Ars Magna, 104–121
Ascent of Rum Doodle, 85
Assyrians, 26
Asteroids, 122–134, 127, 130, 207–224,
    251, 261, 269, 274
Astraios, 52
Astronomical computers, 186–189
Atanasoff-Berry computer, 179
Atkins, P. W., 153
Aubrey holes, 137
Augustine numbers, 82, 259
Augustine, 57, 86, 106, 240
Aztecs, 189, 222

Bab, 175
Baba, Ali, 57
Babbage, Charles, 244
Babbage's engine, 179
Babylonians, 26, 58, 134, 189
Baha'i, 40, 70, 175
Barrow, Isaac, 244
Battles, Ford, 242
Baudelaire, Charles, 205
Beiler, Albert, 10, 36, 73, 94
ben Korrah, Thabet, 97
Berkeley, George, 245

Bernard, St., 185
Bible, 80, 83, 173
Binary numbers, 76, 79, 272
Black Stone, 222–224
Boole, George, 243
Botulis bomb, 223
Bowman, W., 85
Breuil, Abbe, 187
Brouwer, Luitzen, 244
Brown, Frederic, 202
Bruno, Giordano, 106, 157–158
Buddha, 49
Buddhism, 154–157

Calabi-Yau spaces, 196
Calendar, 187, 189
Canchal de Mahoma, 186–189
Cantor, Georg, 200, 204, 244
Carnap, Rudolf, 18
Catenary, 16
Cathedral of Notre Dame, 181, 183–184
Catholics, 57
Cauchy, Louis, 243
Caves, 53, 176–179, 186–189
Cellular automata, 182, 184–185, 266
Celts, 140–142
Chakras, 159
Chomsky, Noam, 205
Christianity, 53, 57, 59, 65, 81, 95, 113,
    148, 185, 220
Christians and Turks, 99–103, 273
Clarke, Arthur, 202, 275
Club of Rome, 68
Cobalt bomb, 223
Cohen, Daniel, 133
Cohen, Henri, 90
Colossus, 179
Comets, 122–134, 207–224, 251, 261,
    269
Complex numbers, 150
Computer code listings, 257–259
Computers, 179, 186–189, 206
Constantine, 99, 103

Constantinople, 101
Constants, 254
Copernicus, 245
Corbusier, 11
Counter-earth, 57
Craters, 215–217, 269, 274
Crowded numbers, 90
Czarnecki transform, 274

da Vinci, Leonardo, 170–171
Dali, Salvador, 171
Dauben, Joseph, 204
Davies, Paul, 45, 104, 149, 167
Davis, Philip, 21, 23, 148, 157
de Acosta, Jose, 165
de Broglie's equation, 18
de Fermat, Pierre, 74
de Sitter, W., 245
Death stars, 122–134, 252
Decimation, 100
Deficient numbers, 87
Descartes, Rene, 243
Descartes'proof of God, 231
Dick, Philip, 203
Dinosaurs, extinction of, 126, 129,
    252
Diophantus, 71
Divine proportion, 168–170
Divine triangles, 74
Doczi, Gyorgy, 167
Dodgson, Charles, 244
Dog star, 189
Donati's comet, 123, 126
Doomsday, 25, 57, 60–61, 66, 126, 207–
    224, 261, 275
Druids, 140–142
Dürer, Albrecht, 76
Durant, Will, 103, 275
Dyson, Freeman, 224

Egyptians, 189
Einstein, Albert, 11, 21, 245–246
Einstein's equations, 16, 18, 81

Empedokles, 51
End-of-the-world, 24–25, 42, 52, 57,
    131, 133, 207–224, 257
Ephesus, 80
Epicurus, 199
Epistle of Barnabas, 65
Epogdoon, 188
Eschaton, 61, 207, 221–224, 248
Estling, Ralph, 25, 99, 104
Ethanomathematics, 166
Euler, Leonhard, 47, 94, 243
Eye of God, 169

Faraday, Michael, 245
Farmer, Philip, 202
Farsi, 175
Female numbers, 36
Fibonacci numbers, 16, 170–171, 265
Fibonacci spirals, 173, 265, 273
Fitzgerald, Edward, 109
Flammarion, Camile, 202
Fractal drums, 120
Fractal quipu, 161–166, 264
Fractal universe, 197
Fractals, 16–17, 119–120, 149–166,
    174, 194, 203, 263
Fractions, continued, 172

Galilei, Galileo, 15
Gardner, Martin, 18, 84, 113, 146,
    171
Gargoyles, 180–186, 266
Gauss, Karl, 38, 81
Gematria, 193
Gnomons, 70, 72
Gnostics, 133
God protection policy argument, 128–
    129, 261
God, the mathematician, 15
Goedel, Kurt, 200, 203–204, 227–229,
    244
Gohonzon, 154–155
Golden ratio, 168–179
Gore, Bernard, 129–130
Gorman, Peter, 51, 53
Gravitation, 15, 16
Gravity's Rainbow, 274
Gregory XI, 145
Grey, Alex, 9
Gutberlet, Constantin, 200, 205

Halley's comet, 220–221
Hardy, Godfrey, 84
Harris, John, 183, 245
Hasidism, 60
Hawkins, Gerald, 139, 141, 176
Haykal, 175
Heraclitus, 11, 29, 48
Hermaphrodite numbers, 56
Hermes, 218–219
Hermetica, 157, 249
Hermetic geometry, 157, 249
Hersh, Reuben, 21, 23, 148, 157
Hexagonal numbers, 272

Hofstadter, Douglas, 302
Hopf bifurcation, 239
Horned spheres, 234
Hoyle, Fred, 15
Hubbard, L. Ron, 202
Hume, D., 143
Hut, Piet, 129
Hypatia, 58–59
Hyperboreans, 53
Hyperspace, 138, 197
Hypertetrahedral numbers, 138
Hypostasize, 192

Iamblichus, 91
Icarus, 218
Imperfect numbers, 91
Incas, 161–166
Infinity, 49, 200–201, 204
Inflation theory, 197
Integer brick problem, 75
Internet and God, 228
Invention generation, 119
Iraq, 189
Iridium, 129
Irrational numbers, 37, 48, 172, 201
Isaac of Nineveh, 122
Islam, 40, 57, 76, 80–81, 95, 102, 148,
    220, 222–224, 274

Jeans, James, 15, 149, 225
Jesus Christ, 53–55, 57, 80, 145
Jones, Inigo, 141, 142
Jouret, Luc, 62, 66
Judaism, 57, 61, 65, 94–95, 134, 148,
    168, 190–193,220, 245
Judgment Day, 25; see also
    Apocalypse
Julia set, 151
Jupiter, 126

Kaaba, 222–224, 274–275
Kabala, 81, 190–197, 274
Kepler, Johannes, 34, 170
Kettering, Charles, 119
Khaldun, Ibm, 97
Kithara, 48–49
Knights Templar, 66
Knuth, Donald, 173, 243
Koch star, 194
Koran, 80
Krieger, S., 93–96
Kronecker, Leopold, 81, 227, 244
Kurzweil Cybernetic Poet, 116, 273

Lactantius, Lucius, 199
Laplace, Pierre, 244
Lee, Siu-Leung, 10
Leibniz' Gottfried, 244
Leibniz' Stepped Reckoner, 179
Light speed, 16, 254
Lintel, 139
Locke, John, 48
Lotus Sutra, 154
Lullian square, 115

Luria, Isaac, 195–196

Madshriti, El, 97
Magic squares, 76
Magnus, Albertus, 95
Mahasaya, Lahiri, 159–160
Male numbers, 36
Malthus, Thomas, 68
Mandala, 155
Mandelbrot set, 150, 152, 203
Mandelbrot, Benoit, 153
Markowsky, George, 171, 173, 230
Marriage between numbers, 36
Marriage in cyberspace, 253
Mathematical Cranks, 254
Mathematical tablets, 26
Mathematicians, religious, 242–244
Matrix prayers, 254
Mayans, 189, 272
Mecca, 222–224
Melencolia, 76–78
Merkava, 191
Mersenne, Marin, 93–94, 243
Meteoroids, 130–131, 222, 274
Millennium, 65
Miller, William, 47
Millerites, 25, 131
Mishna, 195
Misner, Charles, 245
Mithras, 81
Mittag-Leffler, Goesta, 201
Mohammad, 41, 275
Mona Lisa, 171
Montanists, 133
Moon, 187, 215–217
Moore, Patrick, 127
Morris, Henry, 242
Moses, 57
Moslems; see Islam
Muir, Langley, 240
Mutter, Richard, 129
Multiply perfect numbers, 97
Music generation, 116
Musical instruments, 156
Musical intervals, 32, 175

Napier's bones, 179
Narcissistic numbers, 84, 259
Nemesis, 122–134
Newton, Isaac, 15, 18–19, 35, 201,
    243
Nichiren Shoshu, 155
Nichomachus, 85, 92
Niven, Larry, 275
Noah, 57, 80
Norsemen, 132
Notre Dame, 181, 183–184
Number caves, 176–179
Number of the Beast, 71, 75, 79, 174
Number theory, 41, 81
Numbers as gods, 29
Numerical gargoyles, 180–186, 266
Numerology, 41, 193
Numerorum Mysteria, 95

Oblong numbers, 70, 72
Odd numbers, 40
Odd perfect Numbers, 94
Ogdy, 213–214
Ontology, 238
Oort cloud, 127–128
Order of the Solar Temple, 66

Pacioli, Fra, 170
Page, Don, 245
Paley, William, 239
Palindromes, 40
Parthenon, 171
Pascal, Blaise, 242
Pascal's arithmetic machine, 179
Paul, Saint, 132
Pearl–Reed formula, 64, 67, 258
Peecei, Anrelio, 68
Penrose, Roger, 152
Pentaculum, 168
Pentagonal numbers, 45–48
Pentagram, 167
Perfect numbers, multiply, 97
Perfect numbers, 84–98, 260
Phi, 172
Pi swords, 112, 119
Pi, 254, 256
Pielta, La 176–179
Planck's constant, 16, 254
Planets, 24, 34, 37, 56, 254
Plants, 20, 24, 170
Plato, 11, 15, 49, 148, 272
Platonists, 58
Plutarch, 189
Poetry, 107, 115
Polygonal numbers, 47, 258, 272
Population growth, 60–69
Poston, Tim, 233–234, 238
Pournelle, Jerry, 275
Power, Richard, 24
Praying triangles, 73
Prime numbers, 93–94
Proofs of God, 198–206, 227–229
Pynchon, Thomas, 207, 209, 212, 213,
      225, 274
Pythagoras
animals and, 51
barrier of, 188
death of, 49, 57
deformity of, 51
early life, 35,49
infinity and, 201
rules, 35, 53–55, 92
soul, 56
transmigration of souls, 30
Pythagorean triangles, 70, 259
Pythagoreanism, 48
Pythagoreanism, Christianity and, 53–55

Quipu, 161–166, 264

Ra-Shalom, 127
Ragnaroek, 132
Ramanujan Srinivasa, 26, 243
Ramon Lull, 104–121
Raup, David, 129
Reincarnation, 49
Religious mathematicians, 242–244
Revelation, Book of, 25, 71, 134, 187,
      218, 224, 247
Rice, Anne, 60, 186, 255
Riemann, Georg, 243
Rotsler, William, 275
Rubaiyat of Omar Khayyam, 109
Russell, Bertrand, 23, 42, 92, 198

Sadler, William, 146
Sagan, Carl, 9, 198–199, 222, 245
Sahasra chakras, 159
Sapoval, Bernard, 120
Sarsen, 139
Sarton, George, 34
Satan, 256
Satellites, 130
Schimmel, Annemarie, 45, 186
Schroeder, Manfred, 41, 81, 172
Scientology, 202
Seashells, 170
Sefer ha-bahir, 195
Sefer Yetzira, 191
Self-similarity, 17, 150, 170, 174
Sephiroth, 191–192, 274
Seven sleepers, 80
Seventh Day Adventists, 132
Shakespeare, 40, 122, 220, 222
Shekhinah, 192
Shiptonists, 133
Shishdara, 76
Shoemaker–Levy 9, 126
Shri-Yantra, 157
Siberia impact, 126
Sigillum, 168
Sirius, 189
Smith, John Maynard, 251
Smullyan, Raymond, 245
Snow crystals, 19
Sociable numbers, 89
Soka Gakkai, 154–155
Spirals, 18, 169–174
Square numbers, 33, 48, 72
Stapeldon, Olaf, 202
Star of Bethlehem, 202
Star, Koch, 194
Stars, 167–169, 175
Stonehenge, 135–143, 189
Stothers, Richard, 129
String theory, 196
Sufisim, 80
Sumeria, 26, 189
Summa Theologica, 198–199
Swift–Tuttle, 125, 208, 217

Talmud, 195
Tantric literature, 159
Terminology, 247–249
Tetragrammaton, 193
Tetrahedral numbers, 137–138
Tetraktys, 32, 37, 192
Thaumaturge, 48–49, 250
Theomatics, 23, 250
Theory of Everything, 197
Theurgy, 51, 250
Tigris–Euphrates, 189
Tipler, Frank, 203–204, 227, 228
Trachtenberg, 81
Transfinite, 200, 204
Triangles
divine, 74
praying, 73
Pythagorean, 70
Triangular numbers, 32–34, 37–40, 48,
      72, 89, 137, 257
Triangular-square numbers, 38
True Light Church of Christ, 133
Tsunesaburo, Makiguchi, 157
Tunguska impact, 126, 213–214,
      274
Turks and Christians, 99–103, 261,
      273

Ulam, Stanislaw, 271
Umbugio, 42
Urantia, 144–148, 262, 273

Vampire numbers, 255
Vedas, 81, 157
von Braun, Wernher, 245
von Foerster growth, 62, 66, 258
von Hardenberg, Friedrich, 23
von Hoffmannsthal, Hugo, 97
vos Savant, Marilyn, 18
Voting and golden ratio, 173

Wang, Hao, 204, 228
Watts, Alan, 24
Weinberg, Steven, 223
Weissman, Paul, 135
Wells, David, 94
Weyl, Hermann, 15
Whiston, William, 126
Whitehead, Alfred, 244
Witten, Edward, 197
Wronski, Josef, 244

Xenophanes of Colophon, 42

Yates, Frances, 157
Yucatan, 252
Yucca moth, 21

Zohar, 191

About the Author

Clifford A. Pickover is a prolific author and futurist, having published more than 40 books translated into over a dozen languages. Exploring topics ranging from computers and creativity to art, mathematics, parallel universes, Einstein, time travel, alien life, religion, dimethyltryptamine elves, and the nature of human genius, his most recent titles include *Archimedes to Hawking*; *A Beginner's Guide to Immortality*; *The Möbius Strip*; *Sex, Drugs, Einstein, and Elves*; *A Passion for Mathematics*; *Calculus and Pizza*; *The Paradox of God and the Science of Omniscience*; *Surfing Through Hyperspace*; *The Science of Aliens*; and *Time: A Traveler's Guide*. In addition, he has authored more than 200 articles on topics in science, art, and mathematics.

Dr. Pickover received his PhD from Yale University's Department of Molecular Biophysics and Biochemistry, having graduated first in his class from Franklin and Marshall College. Today, he holds over 40 U.S. patents for inventions dealing with computing technologies and interfaces.

Pickover is currently an associate editor for the scientific journal *Computers and Graphics* and is an editorial board member for *Odyssey*, *Leonardo*, and *YLEM*. He also writes the "Brain-Strain" column for *Odyssey*, and his website, www.pickover.com, has received more than a million visits. Dr. Pickover's primary interest is finding new ways to continually expand creativity by melding art, science, mathematics, and other seemingly disparate areas of human endeavor. Other hobbies include Ch'ang-Shih Tai-Chi Ch'uan, Shaolin Kung Fu, and piano. He owns a 110-gallon aquarium filled with Lima shovelnose catfishes, and advises readers to maintain a shovelnose tank in order to foster a sense of mystery in their lives. Look into the fish's eudaemonic eyes, dream of Elysian Fields, and soar.

To reach Dr. Pickover, visit www.pickover.com or write to P.O. Box 549, Millwood, NY 10546-0549 USA.